COR ING

and Core Analysis
Handbook

COR ING

and Core Analysis
Handbook

GENE ANDERSON

Petroleum Publishing Company
Tulsa, 1975

Library of Congress Catalog Card Number: 74–33713
International Standard Book Number: 0–87814–058–1
Printed in U.S.A.

To the wildcatter who gave me a break.
To the pusher who hired me.
To the driller who made me a "hand".

Contents

ACKNOWLEDGEMENT

I would like to thank all who have helped me with advice, critical comment, and material contribution to make production of this book a reality. I especially thank Morris K. Taylor of Christensen Diamond Products, L. Lombos of Robertson Research International, Elliott Moser of Ruska Instrument Corp., E. H. Koepf of Core Laboratories, S.F.L., J. F. Potter of Norwood Technical College (London), N. G. Hortman of Halliburton Manufacturing Services Ltd., M. W. Leighton of Exxon Production Research Co. and D. South of British Petroleum Co. Ltd. Special thanks are given to Jack Davis, formerly with Petroleum International magazine.

Preface

The first coring tool appeared in Holland in 1908 when a driller devised a steel tube which was set in the middle of a drill bit to protect the core from the circulating drilling fluid.

It wasn't until 1921 that J. E. Elliot of the United States made the first effective coring device by combining the toothed bit and the inner core barrel. In 1925 additional improvements were added to include a removable core head, a core catcher, and a stationary inner barrel to which only refinements have been added since.

The aim of this book is to familiarlize the well-site geologist with core-analysis procedures and give him the information necessary to use standard analysis equipment. The book also serves as a reference for those finding it necessary to refresh their memories about sediments, oil properties, coring, and like information.

The sophisticated procedures of full diameter and whole core analysis are not discussed since the equipment required for these analyses is too large and cumbersome for field work. Capillary pressure measurement is discussed but not in great detail because of the time necessary to conduct the procedure.

Field core analysis is important because it measures rock properties when the core is still "alive" with fluids. With plug analysis, reliable data can be accumulated which reflects the in situ recovery conditions.

The single most important variable in the development of correct data is the geologist. His understanding of sediments, their characteristics as regards oil, and gas, and his ability to make uniform measurements are critical to the success of the core-analysis procedure.

In the past, core analysis was not taken seriously and the information many times was considered as "nice-to-know" information. The "nice-to-know" information is now "have-to-know" information if operators are to extract the maximum amount of hydrocarbons from a reservoir. This is increasingly important as consumption of oil increases and outpaces the discovery of new fields.

At the outset, core analysis will appear to be, to the uninitiated, a mysterious, tedious and time-consuming task. With experience, one develops a system to measure the rock sample properties and a "sense" for anticipated, reasonable measurements and calculated results. This sense is not used to juggle figures, but keeps all the figures in perspective and avoids errors. It will be shown that porosity and permeability figures are proportional to one another.

The fluid summation method is not considered because of its inaccuracies when dealing with carbonate sediments. They have usable and nonusable pore spaces; the ratio of these space types is altered with the fluid summation process. To a varying extent, the same ratio is altered in clastic sediments.

It is hoped that the guidelines provided in this book will clarify the more mysterious aspects of the work and will take some of the "grind" out of the routine so necessary for accurate and consistent results. In fact, I hope the presentation of the material will genuinely interest the reader in core analysis.

The Core Analysis Worksheet in Appendix 2 makes it imperative to complete one step before initiating the succeeding step. This makes it virtually impossible to pass a particular step and lose a vital measurement.

The most important advice one may offer to those engaged in core analysis is work methodically and stay organized.

Gene Anderson
London, 1974

1

Important
Properties of
Reservoir Rocks

The estimation of optimum production rates, well spacing, and production allocations have increasingly relied on core analysis data. This trend is responsible for the development of several sophisticated instruments and more refined procedures.

The basic information required to make production estimations are:
1. The effective thickness of the formation.
2. The thickness of the producing zone.
3. The capacity of the reservoir to store fluids.
4. The void space within the rock.
5. The extent to which the void space is occupied by oil and gas.
6. The permeability of the reservoir rock.
7. The position of the gas, oil and water contacts.

The geologist uses the core for detailed lithologic examination and determination of rock age, but he does not use core analysis information *alone* for "making a well." The data evolved, however, serve to set a basis for calibrating the response of wireline devices as well as giving him the best porosity and permeability data possible.

He can be confident of the wireline and (DST) drill-stem test results because he has seen the rock which is the potential producer.

Wireline well logging and drill-stem testing results must be compared with core-analysis results.

The logging geologist should be familiar with the methods of the other

1

evaluation services so he can correlate and assimilate all the information obtained.

1.1 Porosity

Porosity is a measure of the space in a rock not occupied by the solid structure or framework of the rock. It is defined as the fraction of the total bulk volume of the rock not occupied by solids.

Total porosity includes all the pore space in a rock while *effective porosity* includes only the pore space which is interconnected and effective as a void which can be filled or drained of fluids.

To emphasize what porosity means potentially, a porosity of 1% is equivalent to 77.58 bbl of fluid per acre-foot.

A commercial oil-bearing sandstone can have varying porosities depending on the volume of oil(gas) that can be anticipated as producible and the proximity of the potential producer to the consumer. As a general guide, the formation should contain at least 8–10% porosity before it can be considered commercially interesting.

In carbonate reservoirs, this general rule is broken since it is possible to maintain commercial production with as little as 4–6% porosity because of the extremely effective porosity-permeability-fracturing (shear folding) relationship.

Porosity depends to a large extent on the surface texture, roundness, and uniformity of the grain size. The lower the grain sphericity, the higher the porosity. So, spherical grains are not desirable for maximum porosity.

Also important is the cementation and compaction of the grains after deposition. Clay particles, silty cementing material, and shale, when a part of the sedimentary makeup of the reservoir rock, inhibit effective porosity. Clay in its normal wet state is many more times its size than when dry. In such a swollen condition it, along with silt, fills voids and reduces the effective porosity of a potential reservoir rock.

Since field porosity measurements are made on dried samples, the porosity will appear greater than it really is because, previously swollen clays will have shrunk. The loss of porosity due to swollen clay materials can only be measured by reconstitution methods which put the rock sample in its natural environmental conditions where special measurements can be taken.

The individual pore may be like a capillary tube or it may be vugular with druzy crystal infilling; or it may be irregular, feathering out into the bounding constrictions of impermeable cementing materials. The pore may be a thin intercrystalline lenticular opening that is 50–100 or more times as wide as it is thick. The wall of the pore may be composed of clean quartz, chert, calcite, or dolomite or it may be coated with clay mineral particles, accessory platy minerals or country-rock fragments.[2]

The surface area of the rock material surrounding each pore space will change inversely by large amounts as the rock particle sizes become larger or smaller. It is estimated that in a sandstone or a rhombohedrally packed, medium to fine sand, with a particle diameter of 0.01 in. (0.23 mm), there are about 5,000 acres of surface area per acre-foot. This is important in the study and understanding of such reservoir phenomena as wettability, adsorption, capillarity, solubility, and free surface energy.[3]

The crookedness of the pore pattern, called the *tortuosity*, is the ratio of the distance between two points by way of the connected pores to the straight-line distance.[4]

Mathematically speaking the following formulas apply:

$$\text{Pore volume} = \text{bulk volume} - \text{grain volume}$$

$$\text{Percent total porosity} = 100 \times \frac{\text{bulk volume} - \text{grain volume}}{\text{bulk volume}}$$

$$\text{Percent effective porosity} = 100 \times \frac{\text{connected pore volume}}{\text{bulk volume}}$$

In the first case, the grain volume is the determining factor and in the second case, the pore volume is the determinant. Total porosity is not indicative of production potential because some pore spaces may be isolated and not in communication with other spaces. For core analysis, the pore volume as measured in plug or whole core form, connected pore volume, is the effective void in the sample and thus of prime interest.

Pinpoint porosity is a descriptive term used for rocks with tiny isolated pores, visible only with a microscope. This characteristic is typical of chalk and some dolomites and represents a very uniform, highly organized depositional pattern. The size and uniformity of the pore spaces are most easily perceived when filled with oil and viewed under ultraviolet light.

Plates 1–1 to 1–4 show magnified pictures of thin sections of different formations. Comparing the Ordovician Age St. Peter Sandstone with the Pennsylvanian Age Morrow Sandstone, Plate 1–1, one is struck by the similarity of the grain sizes of the two formations and at the same time the lack of similarity of the porosity. The cementing material plays a very important role in the distribution of pore space and consequently (except for fracturing) permeability.

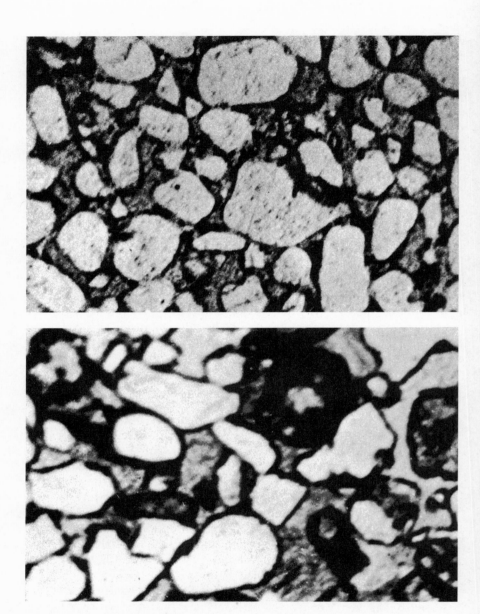

Plate 1-1 Sandstones with similar grain size but different porosities. At top: St. Peter Sand, Ordivician, Webster County, Iowa. Depth 1,200 ft, permeability 2,110 md, porosity 28.4%. Well-rounded grains with very little cementing material. At bottom: Morrow Sand, Pennsylvanian, Beaver, County, Oklahoma. Depth 7,400 ft, porosity 5.0%. Carbonate cement nearly completely fills intergranular spaces. The dark outlines around the quartz grains show carbonate replacement of quartz. Magnification ×50.

The two Oligocene Age Frio Sandstones (Plate 1–2) have very similar porosity but absolutely dissimilar permeabilities. The dissimilarity is caused by the difference in sorting and amount of matrix material.

Plate 1–3 compares two sandstones having large differences in both porosity and permeability. Plate 1–4 shows the wide variation of permeability in carbonates.

Another formula often used in the laboratory for determining grain volume is:

$$\text{Grain Volume} = \frac{\text{dry sample weight} - \text{submerged sample weight}}{\text{density of saturant}}$$

This formula is difficult to apply in the field. Errors develop because pore spaces will trap air to a degree dependent on the orientation and size of the individual grains.

Microscopic analysis of core samples should be done *before* as well as *after* the sample has been cleaned with a solvent. Sometimes the cleaning action increases the pore space by destroying the natural surfaces of the rock. For example, some dolomites with pinpoint pore space are filled with halite. Water placed on these samples will drastically alter the porosity of the sample.

1.2 Grain Orientation

Problems in core analysis develop when cross-bedding and graded bedding are encountered. The frequency of sampling should be adjusted to accommodate any sedimentary changes and variations. The orientation of grains in any rock matrix helps determine the size and orientation of pore spaces in clastic sediments.

Carbonate sediments pose a separate problem in establishing the cause for and size of the pore-space pattern since chemical solutions and fracturing (Fig. 1–1) play such important roles in carbonate constitution and the common-place recrystallization and alteration of the rock. Fig. 1–2 shows various types of limestone porosity.

Clastic sediments are deposited so that the grains are oriented in the position of greatest stability. This means that the flaky grains found in shale will lie parallel to the bedding plane and angular or semirounded grains will lie parallel to the current depositing the sediment.

Normally the larger, heavier ends of the grains will point upstream and be wedged between other grains. This is an attempt by the grains to stabilize between the vectors of gravity and motion. As a result, hori-

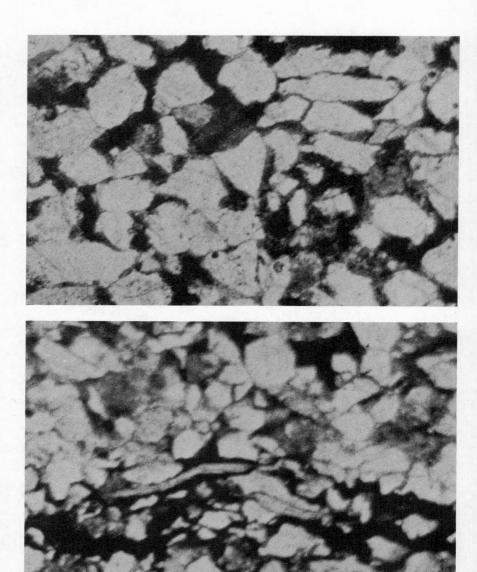

Plate 1-2 Two sands of some age have similar porosities but dissimilar permeabilities. At top: Frio Sandstone, Oligocene Age, South Louisiana. Permeability 2,100 md; porosity 28.5%; depth 11,500 ft. Good sorting of subangular grains, clean pores with very little cementing material. Magnification ×50. At bottom: Frio Sandstone, Oligocene Age, South Louisiana. Permeability 160 md; porosity 25.8%; depth 11,480 ft. Moderate sorting with a thin shale streak with aligned mica flakes. Some intergranular brown silt and clay size minerals are present. Magnification ×50.

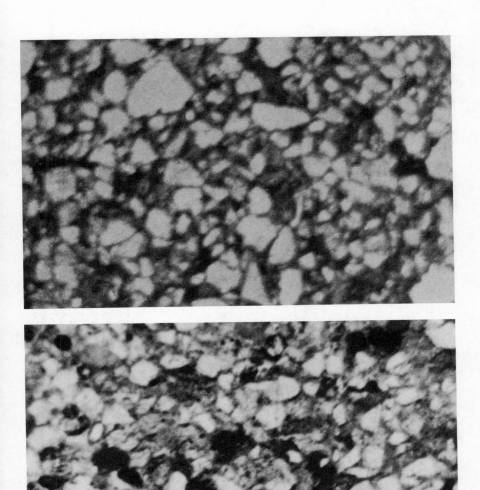

Plate 1-3 Two sandstones having large differences in both porosity and permeability. At top: Frio Sandstone, Oligocene Age, South Louisiana. Permeability 0.3 md; porosity 13.9%; depth 11,450 ft. Poorly sorted sandstone of angular grains. Much intergranular brown silt and clay size minerals are present. Magnification ×50. At bottom: Navarro Sandstone, Upper Cretaceous Age, Frio County, Texas. Permeability 11 md; porosity 30.4%; depth 3,100 ft. Large amounts of brown silt and clay size minerals are present. Much of the clay is of the swelling types. The pores are very small. Magnification ×50.

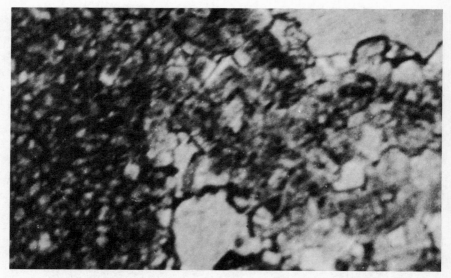

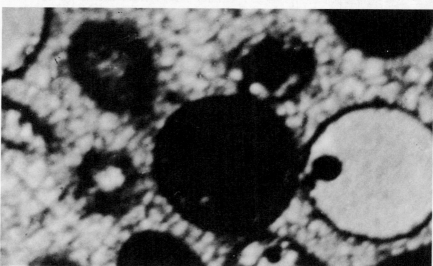

Plate 1-4 Two carbonates showing wide variations in permeability for similar porosity. At top: Trempealeau Formation, Cambrian Age, Morrow, Ohio. Permeability 840 md; porosity 11.8%; depth 3,000 ft. Dolomite with solution cavities. Less conspicuous intercrystalline pores are also present, giving the formation a great variation in pore sizes and geometry. Magnification ×50. At bottom: Lansing-Kansas City Formation, Pennsylvanian Age, Barton, Kansas. Permeability 0.05 md; porosity 10.1%; depth 3,350 ft. Oolitic and oolicastic limestone showing fair porosity with very low permeability. Magnification ×50.

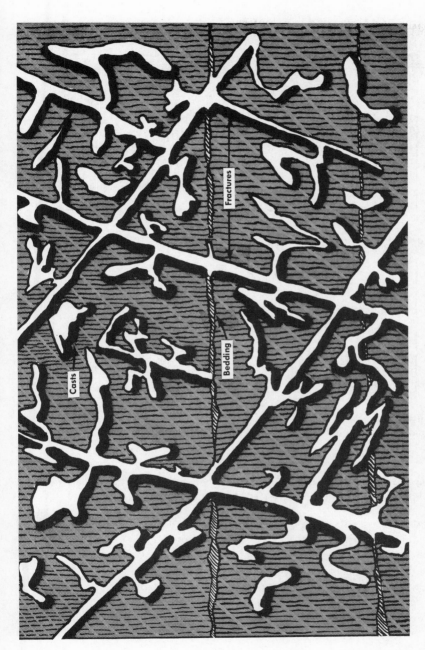

Figure 1–1 Idealized section showing how fractures aid in increasing permeability. Solutions passing through dissolve the wall material and widen the fractures, which connect otherwise isolated cavities and pores. From Levorsen, A. I., "Geology of Petroleum," second edition.

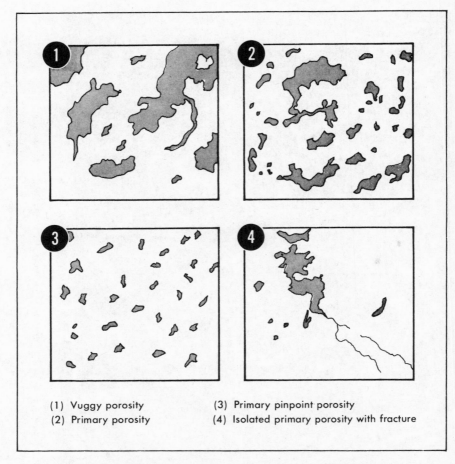

(1) Vuggy porosity   (3) Primary pinpoint porosity
(2) Primary porosity   (4) Isolated primary porosity with fracture

Figure 1-2 Typical limestone porosities: (1) vuggy porosity, (2) primary porosity, (3) primary pinpoint porosity and (4) isolated primary porosity with fracture.

zontal permeability tends to be greater and more uniform than vertical permeability with respect to the bedding plane.[6]

More subtle are the changes in a rock's porosity and permeability as caused by folding, faulting and plastic deformation. Two basic types of tectonism affect a rock's acceptability and transmissibility of fluids — shearing and flexure folding.

Shearing is most common in carbonates and takes the form of vertical fracturing from compression, tension and torsion. Carbonates have little in the way of weak bedding planes which allow bending of the formation.

Flexure folding, or common folding, occurs normally in clastics which show bedding planes as the weak aspect of the formation since they are the slip planes or glide planes along which the formation bends or flexes.

Faulting, fracturing, and folding upset the balance of rock-fluid relationships established during deposition. They indirectly cause the sedimentary fabric to change through the movement of blocks and the shifting of grain axes. This helps induce or inhibit the migration of interstitial fluids. From the stress and strain placed on the sediments, recrystallization and sometimes granulation takes place.

Solutions in pore spaces control to a large extent the stress-strain and fracture relationships found in sediments. Experiments in limestone show that ductile formations become more brittle as the interstitial pore fluid pressure is increased in relation to the confining grain pressure. Pore pressure greater than 0.47 psi/ft (pressure gradient), is considered abnormal. The highest limit is 1.0 psi/ft—the weight of the overburden—which represents a formation specific gravity of 2.6 with a water-filled porosity of around 20%. The 1.0-psi/ft gradient can be exceeded considerably where the formation has been tectonically shifted, such as the limestones and dolomites in Kentucky. Here the pore pressure can be as high as 1.5 psi/ft which is enough to cause formation fracturing.

Hubert and Willis have shown that in areas which have already gone through tectonism, fracturing of a formation was related to the pore pressure by the general formula:

$$P_f = D/3 \ (1 + 2p)$$

where: P_f = fracture pressure, psi
   D = depth, ft
   1 = 1.0 psi/ft, overburden as a gradient
   p = formation pressure, psi/ft, as a gradient.

Within this framework, fracture pressure equals the pore pressure (formation pressure) at overburden pressure (1.0 psi/ft). Fig. 1–3 shows the force-deformation curves for the Indiana Limestone. The points a, b, c, d, and e are yield points at which fracturing takes place. The grain yield strength increases as the pore pressure decreases. The formation fails by brittle fracture when the pore pressure equals the confining grain pressure. When the pore pressure is reduced somewhat, several long shear planes develop instead of one brittle fracture and with no pore pressure and the same confining pressure, a large number of short shear planes develop.

The effective pressure of the grains (confining pressure) minus the

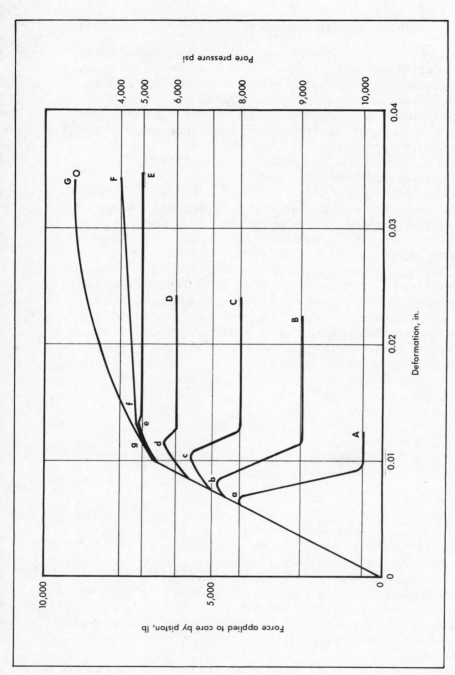

Figure 1-3 Force-deformation (stress-strain) curves for Indiana limestone under varying values of pore (interstitial) pressure. All tests run at confining pressure of 10,-000 psi at room temperature. Note the greater ductility at lower pore pressures. Reprinted by permission from Colorado School of Mines Quarterly, Volume 54, No. 3, Copyright © 1959 by the Colorado School of Mines.

pore pressure, controls the amount of fracture facility. While the prospect of fracturing looms as a possibility, one is confronted with the likelyhood of grain alteration. Grain alteration depends on temperature, orientation of crystal axes, strength of each crystal type's structure, hardness of the crystal faces, and how the crystals are interlocked. All these factors will contribute to relieving pressure by reorientation or secondary mineralization.

In the very prolific Asmari limestone of southwest Persia, oil is produced from fractures. The fracturing is prominent along the pitch of the folds. Halite overlies the Asmari and therefore seals the oil zone, prohibiting further vertical migration. Recrystallization of the Asmari was induced by, or happened in conjunction with, fracturing and faulting. The result is that, generally, the porosity has increased from 2 to 15% and the permeability has increased from 0.00005 to 0.5 md.[9]

In California, oil is produced at Santa Maria from fractured shale and sands. At Monterey, oil is produced from fractured cherts which are interbedded with brittle sandstones and shales.[10]

Plastic deformation of rock grains normally accompanies folding and faulting. Calcite crystals may form secondarily or deform into dolomite rhombs. Round oolites can deform into ellipsoids. Halite and anhydrite are notorious for their zones of deformation.

The constituent elements of pore fluids play an important role through ionic exchange with rock cementing material and weak crystal structures. This ionic exchange can increase or decrease the effectiveness of the rock as a reservoir.

Permeability along the strike of a fold can differ markedly from that along the dip due to the grain orientation, displacement, and the addition or loss of cementing material through solution. On a smaller scale it may differ along the strike and dip of a core. The direction of the highest permeability in a core is a factor which requires careful consideration when proceeding with an analysis. Ideally, the sample plugs taken from the core should be cut parallel to the bedding plane and perpendicular to the strike of the fold or structure (down dip). To determine which way is "down-dip" on most cores is virtually impossible on any decent-sized structure unless the drilling has been along the flank.

Fig. 1–4 shows the relationship between porosity and permeability. This relationship, with only slight variations, is representative of most sandstones. It does not represent a porosity-permeability relationship for carbonates since they have a haphazard pattern of porosity and permeability.

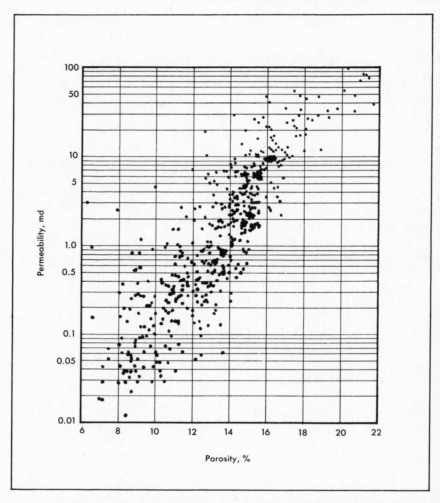

Figure 1–4 Results of porosity-permeability measurements from about 500 samples of the Middle Devonian Bradford sandstone in northwestern Pennsylvania. Although this sandstone is considered very uniform there is not a compact porosity-permeability trendline. It is probable that both the cementing material changes and the uniformity of sample measurement contribute to such wide scatter. From Ryder, World Oil, May 1948 p. 174.

1.3 Permeability

Permeability is a measure of a rock's ability, under a potential gradient, to conduct fluids. This "measure" represents the avenues or communication lines between pore spaces.

Permeability is not a measure of void space as often erroneously expressed. Like porosity, however, it depends on rock properties such as grain shapes, grain and cement texture, angularity, and size distribution.

Unlike porosity which is theoretically independent of the actual size of the rock grains, permeability is very much dependent on rock grain size (rock grain size should not be confused with rock-size distribution). The smaller the grains, the larger will be the surface area exposed to the flowing fluid. The surface area creates a drag on the fluids which limits the flow rate.

Liquids flow faster in the center of the pore than along the sides. This, a differential fluid flow, is called the "Klinkenberg Effect." Thus a small grain size has the effect of lowering the permeability.

The best example of this principle is a shale which abounds in pore space, yet has almost no permeability. It follows that dirty, silty sands have lower permeability than clean sands.

The unit of measurement used to quantitatively express permeability is the Darcy, named after Henri Darcy who experimented in the 1850s with the passage of fluids through porous media. Using any given value for permeability noted as "k," the flow rate through any porous rock sample is proportional to the difference in pressure across the sample, the viscosity of the fluid, and the length of the sample through which it passes. The American Petroleum Institute has modified the darcy to standardize its use for the petroleum industry. "A porous medium has a permeability of 1.0 darcy when a single-phase fluid of 1.0 cp viscosity that completely fills the voids of the medium (rock sample) will flow through it under conditions of viscous flow at a rate of one $1 cm/sec/cm^2$ cross-sectional area under a pressure or equivalent hydraulic gradient of one atmosphere (76.0 cm of Hg)/cm."

Darcy units are too large to be convenient for our use because many rocks have less than 1.0 darcy permeability so the millidarcy (md) is used — 0.001 darcy. In Table 1–1 a few permeabilities are shown as they relate to porosity.[9]

Mathematically, Darcy's Law is expressed:

$$\frac{Q}{A} = \frac{k}{\mu} \frac{\Delta P}{L}$$

where;　Q = flow rate, cc/sec
　　　　A = cross sectional area, cm²
　　　　L = length, cm
　　　　μ = viscosity of flowing fluid, cp
　　　　ΔP = pressure differential across sample, atm
　　　　k = permeability, darcys

Although field core analysis uses air to measure permeability, one must be cognizant of a very important fact first discovered and experimented with by Klinkenberg. He discovered that permeability measurements made with air as the flowing fluid showed different results from permeability measurements made with a liquid as the flowing liquid.

Klinkenberg also found that for a given porous medium, as the mean pressure increased, the calculated permeability decreased. This lends some credence to the time proven expression "high pressure — low volume, low pressure — high volume." Every field has its own unique characteristics and today's more sophisticated analytical methods qualify better the reasons for certain reservoir pressure and volume phenomena — but never quite so directly in understandable language.

TABLE 1-1 POROSITIES AND PERMEABILITIES OF VARIOUS OIL RESERVOIRS[13]

Pool	Porosity		Permeability	
Location, reservoir	Range %	Average %	Range md	Average md
12 pools producing from Smackover (Jurassic) limestone in southern Arkansas	12.5–21.3	16.9	50–2,000	737
Masjid-i-Sulaiman oilfield, Iran, Asmari limestone		2		0.0005
		5		0.007
		10		0.05
		15		0.5
Rangelyfield, Colorado, Weber sandstone (Pennsylvania N)		16		20
East Texas pool, Texas, Woodbine sand (Upper Cretaceous)		25	Up to 4,600	1,500
Ten Sections, Kern Co., California, Stevens sand (Upper Miocene)	15–30	20	10–3,000	140

As mentioned, permeability measurements are made with dried samples and as in porosity measurements, clay particles or flakes in the reservoir rock influence the ability of a rock to allow fluids to pass through it. Permeability of fresh-water sands when clay is part of the matrix is considerably less than that found with salt-water sands or dried sand samples (Fig. 1–5). Salt-water permeabilities according to Johnston and Beeson are probably the most representative.

To measure the permeability most accurately one must use the same fluid that exists in the reservoir rock as a measuring fluid. This is not practical in field analysis because gas, oil, and water are present and flow in mutual contact with one another where a pressure gradient exists.

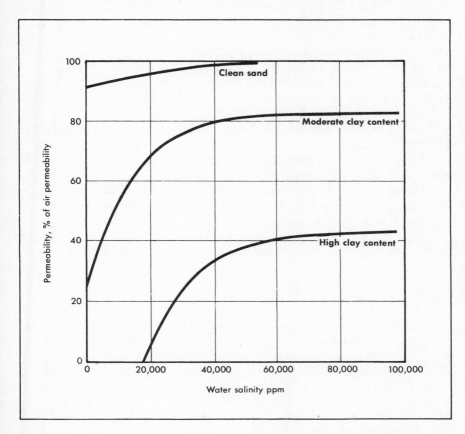

Figure 1–5 Variation in water permeability with changing salinity and clay content. From The Fundamentals of Core Analysis, Core Laboratories Inc., p. 46.

Each of these is capable of flowing through a given reservoir rock with a different degree of facility inversely proportional to the viscosity of the flowing medium.

Further experimentation by Wyckoff and Botset has shown that the relative amounts of each fluid flowing in a multiphase system depends not only on the viscosity of the flowing medium, but also upon the relative fluid saturation of the rock at any particular point within the rock. These sophisticated analyses must be left for the laboratory where full diameter and reconstruction core analyses can be accomplished.

Permeability normally varies from one location to another vertically as well as horizontally in a reservoir rock. As already shown in Table 1–1 there are fairly wide variations in the same reservoir formation and even wider variations in the formation of one area to the same formation in another area. What is noticeably similar is the ratio of porosity to permeability in each specific formation. This ratio forms a trend line as shown in Fig. 1–4.

The causes are numerous for any wide variation in the porosity/permeability ratio. Chief among the causes are sediment sorting, fluid migration, and cementation. Vertical permeability is nearly always less than horizontal permeability except where shear folding is prominent as in carbonate structures.

Where vertical permeability is greater, it is associated with a certain amount of compression and torsion fracturing.

Solutions, forced to migrate because of structural dynamics and inordinate pressure concentrations, leach out cement and/or foster primary and secondary crystallization.

High vertical permeability can be identified on microlaterlogs by the negative separation of the curves.

1.4 Saturation

Ideally, a petroleum reservoir is formed in such a way that the fluids in it are separated simply by the force of gravity. Thus from the top to the bottom of the reservoir there will be gas, oil, and water.

In the practical case, gravity plays a large part but nearly always, water is found in the oil zone as a thin film on rock grains and in more inaccessible capillary pores which cannot accept invasion of the more viscous oil. Such water is classified as connate water; it is present before the oil migrates into the reservoir. The connate water may amount to as much as 50% of the fluids in the reservoir.

The greater adhesion of water over oil to rock material prevents the oil from displacing the grain-surface water and the capillary-pore water. Nevertheless, the masses of oil and water are separated into zones by gravity and saturation analyses are important to delineate their mutual boundary.

Variable oil saturation exists in some reservoirs where the separation of gas, oil, and water does not depend on gravity separation but on impervious rock material. For example, there are potential reservoirs where the oil has been found with completely water-wet layers above and below in the same apparent formation material. Closer examination shows that thin algal layers create an impermeable membrane which separates the fluids.

The more unusual saturation variations are found in carbonate rocks and each case must be explained on its own merits within the framework of sedimentation, structure, paleontology, and the individual geologist's knowledge and experience. The ultraviolet light is indispensable for the geologist since it assists in determining not only the probable gravity of the oil but the pattern of oil in the rock, which is the pore pattern itself.

In complex reservoirs, the water zones normally have lower permeabilities than the oil zones, especially in sedimentary traps.

Saturations determined with routine core analyses reflect very poorly the quantitative conditions of fluid and gas at depth. The reservoir rock when taken as a core is flushed during drilling; it is subjected to a severe pressure and temperature change when brought to the surface which fosters rapid fluid evaporation. Only with a pressurized core barrel can actual saturations be determined. This type barrel is rarely used because of the expense, and there are simpler ways to develop the necessary saturation values.

Table 1–2 shows that the measured values of gas, oil, and water cannot be relied on as proportional to the actual reservoir fluids at depth.

TABLE 1–2 TYPICAL SATURATIONS IN A CORING OPERATION

	S_o	S_w	S_g
Fluid percentages of undisturbed reservoir formation.	0.59	0.41	0.0
Fluid percentages of core in the barrel at the bottom of the hole.	0.16	0.84	0.0
Fluid percentages in core at the surface.	0.13	0.40	0.47

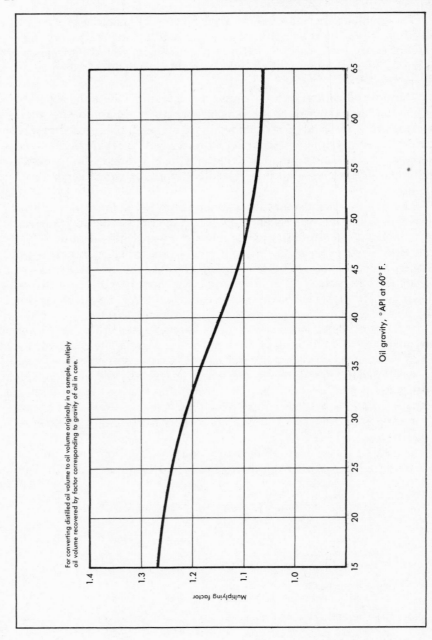

For converting distilled oil volume to oil volume originally in a sample, multiply oil volume recovered by factor corresponding to gravity of oil in core.

Oil gravity, ° API at 60° F.

Multiplying factor

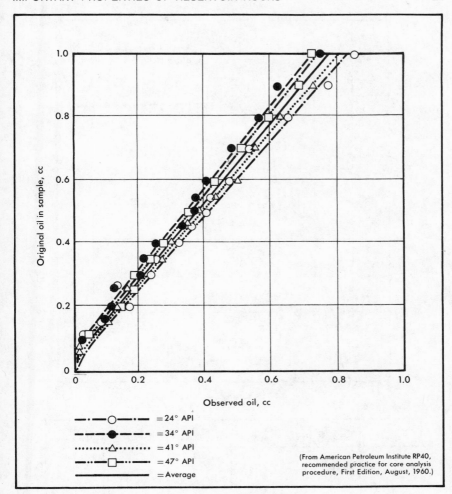

Figure 1-6 Distilled oil volume-correction factor.

Thus, the Fluid Summation Method of analysis must be looked upon with a very skeptical eye.

When oil saturation in a core is determined by the retort method (covered in Chapter 6), the amount of oil collected is not a true reflection of the amount of oil in the core. Retorting causes oil to crack and coke, and the fluid collected is of higher gravity and lower volume than the original fluid. Oil gravity plays a part in this process, and Figs. 1–6A and 1–6B

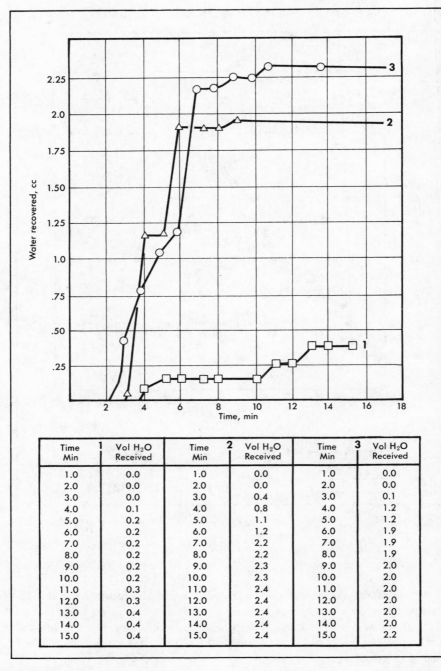

Time Min	1 Vol H$_2$O Received	Time Min	2 Vol H$_2$O Received	Time Min	3 Vol H$_2$O Received
1.0	0.0	1.0	0.0	1.0	0.0
2.0	0.0	2.0	0.0	2.0	0.0
3.0	0.0	3.0	0.4	3.0	0.1
4.0	0.1	4.0	0.8	4.0	1.2
5.0	0.2	5.0	1.1	5.0	1.2
6.0	0.2	6.0	1.2	6.0	1.9
7.0	0.2	7.0	2.2	7.0	1.9
8.0	0.2	8.0	2.2	8.0	1.9
9.0	0.2	9.0	2.3	9.0	2.0
10.0	0.2	10.0	2.3	10.0	2.0
11.0	0.3	11.0	2.4	11.0	2.0
12.0	0.3	12.0	2.4	12.0	2.0
13.0	0.4	13.0	2.4	13.0	2.0
14.0	0.4	14.0	2.4	14.0	2.0
15.0	0.4	15.0	2.4	15.0	2.2

Figure 1–7 Tables and graph of water saturation characteristics in limestone, Persian Gulf.

show the expected difference in the amount of oil in the sample and the amount collected in the retort, as a function of oil gravity. The gravity of the oil collected is *not* the gravity of the oil in the reservoir.

Figure 1–7 shows in tabular form three cases of water saturation and their corresponding graphical curves. These graphs were developed from limestone cores.

1.5   Relative Permeability

What has been described under the heading "permeability" is absolute permeability measured using air which is much less viscous than the fluids found in any reservoir.

Specific permeability is defined as permeability with only one fluid phase present. If one was to measure permeability with the actual fluids found in a reservoir, very different results would be recorded. These different results are termed effective permeability of the particular mobile fluid phase.

The ratio of effective permeability to specific permeability is the relative permeability.

Relative permeability is the ability of a rock to allow fluid to pass through it under a potential gradient when there are two or more phases such as oil, gas, and water present in the rock's pore space. How much and which phase fluid will be produced depends on the permeability of the rock and on the degree of saturation of each of the fluid phases.

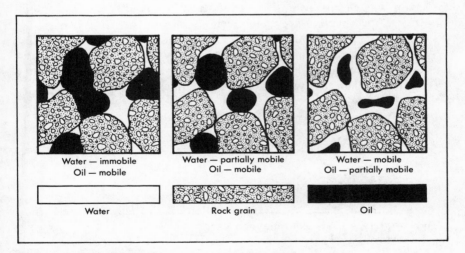

Figure 1–8 Various conditions of saturation.

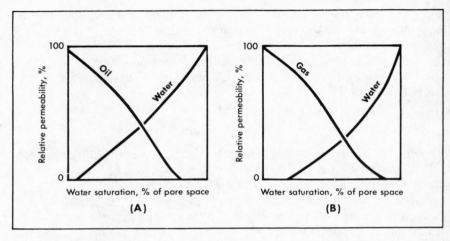

Figure 1–9 Examples of relative permeability curves: (A) for rock containing oil and water; (B) for rock containing gas and water.

Sophisticated laboratory equipment is necessary to conduct tests with regard to relative permeability. Since the required measurements are not ordinarily taken with a "live" core at the well but are taken with a "dead" core after several days of transport to the laboratory, the actual results as they apply to a specific well must be viewed with caution. Experimentation with the fluid phases and the holding reservoir rock are essential to understanding the mechanics of the reservoir but great care must be taken when transposing lab results from a "dead" core to the fluid phases of a prospective producing well.

Fig. 1–8 shows various conditions of saturation that might be found in any given reservoir. Fig. 1–9 shows oil and water, and gas and water curves with respect to relative permeability. Fig. 1–10 depicts the relative permeability when the two phases are oil and gas.

Figure 1–10 Typical gas-oil relative permeability curve. From The Fundamentals of Core Analysis, Core Laboratories Inc., p 55.

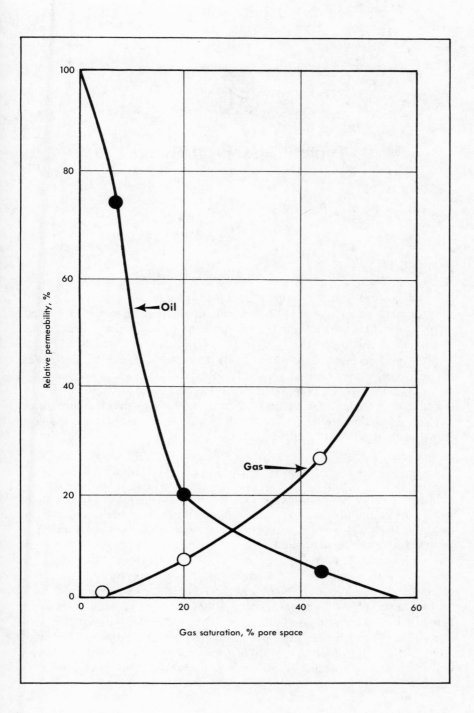

Typical Reservoir Rocks

Oil and gas are found in several sedimentary and other provinces. The environment plays a very important role in whether oil or gas is formed. (See Appendix C for a detailed classification of sediments based on environmental conditions.)

It would be impossible in this work to discuss all the provinces and therefore the main reservoirs — sandstone, limestone, and dolomite — will be considered since it is in these rocks that one normally finds oil or gas. Almost as a reflex, the reservoir rocks have been divided into three classes, as just mentioned. Actually they will be treated in two classifications: carbonates and clastics.

2.1 Carbonates

Limestones, and dolomites which are technically part of the limestone group, have unique structure and composition. The rocks are unique and reflect vividly the environmental, chemical and biological conditions in which they were deposited. Their classification is determined by any one of several methods.

There is no single "best" classification method for carbonates. Quite often the terminology developed and used during a well program depends on the nature of the carbonates being drilled and the reference text being consulted. Ordinary logging of cuttings while drilling, without thin sections, is a difficult process through which to chart the history of the sediments. More often than not, without the services of a paleontologist and thin sections, mud logging is used only to differentiate one rock type from another.

When confronted with a full core, one has the whole evidence and is required to develop in greater detail — as much as time, sophistication of examining equipment, and knowledge will permit — the history of the cored section.

Two major classifications of carbonate rocks are:

1. *Descriptive:* chemical, biological, mineralogical, physical and textural.

2. *Genetic:* chemically precipitated limestones, fore-reef talus limestones, fecal pellet limestones, hydrothermal dolomite.

These two classifications are often used in conjunction with one another to make interpretations of sedimentation more meaningful. For the geological engineer, an easily applied classification system which can be rapidly applied is essential. The system must basically allow for a clear description of the sample which can be expanded as more detailed information is developed from closer examination.

To develop more detailed information core analysis is best conducted by a team with two or three geologists. With a team effort the necessary information can be developed and compiled in an orderly fashion while the ordinary logging operation can be continued correctly when the bit gets back on bottom.

Fig. 2–1 depicts the basic organization of carbonate formation and development. The essential minerals of carbonates ($CaCO_3$) are calcite, aragonite and dolomite. Vaterite, also a carbonate, is so unstable it changes to a calcite in a very few days. Aragonite is also unstable but takes years to change to the more stable calcite.

Most dolomites are the result of post-depositional processes and usually show some relationship with calcite. When silica material is present with a preponderance of carbonate, it generally takes the form of chert or chalcedony.

Accretionary and biochemical limestones are considered the primary carbonates since they form directly by the extraction of $CaCO_3$ from seawater through inorganic or organic processes. A minor amount of rock develops as a direct result of freshwater chemical precipitation. Such accretions are tufa and travertine. Shallow marine waters, partially or intermittently enclosed, may become saturated enough to precipitate aragonite. Biochemically formed accretions have either a biohermal or biostromal character.

Biohermal limestones have been described as dome-like, mound-like and lense-like. They are built exclusively or mainly of organisms and are enclosed in a rock of a different lithologic nature. Whether or not a bioherm is a reef or not depends on its wave resistance potential. A reef is

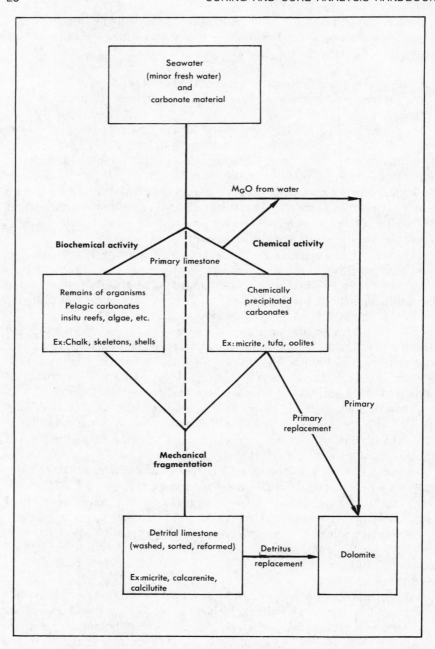

Figure 2–1 Classification of carbonate rocks.

described as "a skeletal limestone deposit formed by organisms possessing the ecological potential to erect a rigid, wave-resistant, topographic structure."[18]

Biostromes, formed by sedentary organisms such as crinoids, shells, and algae are stratiform over large areas in conjunction with other rocks. Algae often separates deposition cycles and consequently are important to the porosity, permeability, fluid migration control barriers, and pressure anomalies in contiguous rock units.

Some biochemical limestones are the products of floating (pelagic) organisms which, being microscopic, leave very fine-grained tests on the sea-floor. Foraminifera are chiefly responsible for pelagic deposits and where found must be Cretaceous or younger since "forams" did not develop the lime-secreting habit until that time.

Chalk is described by G. E. Thomas[19] as ". . . porous, microtextured, and friable variety of carbonate rock. Partly of chemical origin, but also represents the 'flour' formed by disintegration and abrasion of skeletal, nonskeletal grains, and algal growths. Chalky material referred to in this study is microtextured (0.01 mm approximately), has calculated porosities up to 30%, high connate water content, and virtually no oil saturation because of fine capillary pores. Generally deposited in the quiet-water environments of lagoons or intershoal areas of carbonate shelves."

It is made chiefly of calcareous tests of plankton and is Cretaceous mainly in age. Chalk is rarely altered in spite of the ability of solution to move through it. This stability is thought to derive from the fact that chalk is precipitated as a calcite, the most stable of the carbonates.

Micrite varies from lime mud ooze to solidified rock made predominantly of calcite crystals. It has an upper size limit of 0.03 mm. The term micrite is used to describe rock particles which are mud-like in texture. It will be hard to differentiate calcilutites from micrite because of the similar nature of the materials. In fact, the name micrite is often used as a general description for both primary and detrital sediments. Only the associated materials will yield the information with which to form an appropriate opinion as to the rock type.

Micrite is "microclastic" and "microcrystalline." Microcrystalline types have a high luster and are characterized by a mosaic of tightly interlocking calcite crystals. Microcrystalline micrite is commonly associated with micro-grained dolomites and anhydrite. Microclastic micrite is characterized by dull luster, occasional fossil fragments, silt-sized quartz grains, and perhaps a few calcite grains. Some chalks may belong to this group.[21]

Primary limestone sediments provide the bulk material for the mechanically transported detrital limestones. They may be grossly differentiated in the following way:

Primary limestone	*Detrital limestones*
1. Quiet water	1. Various degrees of turbulent water
2. Shale, clay, mud, lime ooze	2. Sandstones, orthoquartzites, clear calcite cement assoc.
3. No size sorting	3. Size sorting
4. Fossils articulated	4. Fossils disarticulated
5. No current bedding	5. Current bedding

Carbonate detritus is subaqueous in origin. Sedimentary grains up to 2 mm in diameter formed by mechanical deposition are termed calcarenites. Those which are larger than 2 mm are termed calcirudites. The composition of the "grains" covers a wide variety of debris.

Calcarenites are current-bedded detrital limestones with common cross-bedding. They are formed primarily from chemically precipitated limestone such as oolites and aragonite muds. Calcilutites are calcarenites of micritic size and are really consolidated muds. These consolidated muds show a conchoidal or subconchoidal fracture capability and because they were the best quality stone used in lithography, they are often termed lithographic limestones.

The top and bottom of beds can be determined by knowing the living habits of benthonic (bottom dwelling) organisms which are buried with the skeleton intact and where the filling sedimentation takes place in quiet waters. Stromatolites, excellently disposed for such marking and commonly called "algal growths," range in age from Precambrian to Recent sediments. They vary in size from a few millimeters in breadth and height to tens of meters. The internal layers of stromatolites are convex as one looks up the column.

Photomicrographs of various limestone types are shown in Plate 2-1.

The name dolomite will, it seems, always apply to the mineral as well as the rock. Where up to 90% of the rock is made of dolomite, the rock is so named. If only between 50–90% is dolomite, it is called a calcareous dolomite. If the dolomite percentage is below 50%, the rock is called a dolomitic limestone until there is no dolomite present in the rock. There are no even gradations from limestone to dolomite and a carbonate is normally close to being all one or the other.

The chemical formula which expresses the replacement process is:

$$2CaCO_3 + MgCl_2 \rightarrow CaMg(CO_3)_2 + CaCl_2$$

Dolomitization is the metasomatic process of chemical change which, in varying degrees, alters the original texture and crystallization of the rock. Dolomites are more uniformly grained than limestones; they are devoid of fossils; the rhombs are commonly zoned with alternate clear and clouded zones, and they can grade laterally into limestone.

There are still many unknown details and puzzling features about dolomites and their relationship with other carbonates but generally the provenance (provenience) of dolomite is replacement of limestone. Perhaps the strongest evidence that dolomites originate as replacement material is the fact that even though no known organisms secrete dolomite, whole coquinoid beds are now all dolomite.[22] The so-called primary dolomites are usually finely laminated and are thought to develop from water more saline than seawater but not saline enough to precipitate sulfates and chlorides.[23]

There is difference of opinion about the change in porosity that occurs when a limestone is replaced by a dolomite. If replacement takes place on a molecule-for-molecule basis, the change in volume from limestone to dolomite amounts to a shrinkage of 12%. This change should leave some new void space and generally speaking this theory is borne out since dolomite reservoirs are more prolific oil producers than limestone reservoirs. According to Hohlt in "The Nature and Origin of Limestone Porosity," petrographic studies show that in limestone, calcite crystals tend to orient their C-axes in the bedding plane. In dolomites the C-axes are randomly oriented. He explains this by saying that the shrinkage of the C-axes as the limestones become dolomites creates loose crystal packing, allowing movement of the axes and consequently more void space.[24]

Mottled dolomite is the result of incomplete dolomitization and exhibits itself as porphyroblasts in an altered calcareous matrix or as scattered patches of dolomite.

Photomicrographs of various dolomites are shown in Plate 2–2.

Solutions have a great effect on limestones and cause the abundance of dolomite to be formed. The solutions often move in limestone along stylolitic seams, through which a surprisingly great deal of material can be moved. It always seems that reservoir carbonates do not have uniform evenly distributed porosity and permeability. Much of this phenomenon can be attributed to the uneven distribution of dissolvable rock material and the composition of erosion-causing solutions.

Variegated porosity through a reservoir is thus the rule rather than the exception. Selective solutions (those of a particular pH) may remove fossils or some particular part of the matrix. Generally, fractured and vuggy carbonates seldom have high porosities but the permeabilities are very often enormous.

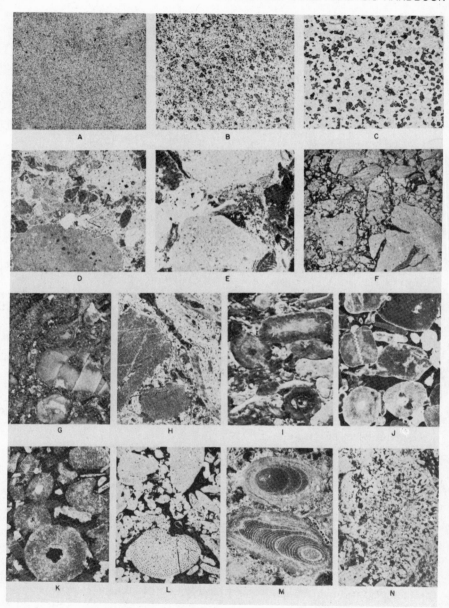

Plate 2-1 Limestone rock types.

Explanation Plate 2-1.   Limestone rock types[28]

A. Micritic limestone (microclastic type). Dominantly micrite with few quartz silt grains. Mississippian Lodgepole, Gallatin Co., Montana, ×8.

B. Micritic limestone (microcrystalline type). Composed of interlocking mosaic of calcite crystals 0.03 mm diameter. Mississippian Charles, Hot Springs Co., Wyoming, ×8.

C. Dolomitic micritic limestone. Dolomite rhombs scattered in matrix of micrite. Jurassic Arab, Saudi Arabia, ×8.

D. Conglomeratic-micritic limestone. Composed of subangular to subrounded micritic limestone fragments with a calcareous clay matrix. Triassic limestone, Sicily, ×8.

E. Limestone conglomerate. Composed of subrounded to rounded limestone rock fragments and fossil debris. Miocene age, Philippines, ×8.

F. Detrital limestone (with intraclasts). Contains fragments of pellet and micritic limestone derived from nearby, partially consolidated bed. Pennsylvanian Hermosa, San Juan Co., Utah, ×8.

G. Skeletal-micritic limestone. Contains fragments of crinoids, mollusks, bryozoa, and corals in a micritic matrix. Pennsylvanian Hermosa, San Juan Co., Utah. Photograph of an etched sample, ×8.

H. Same as G, but a thin section photograph instead of an etched sample photo, ×8.

I. Skeletal limestone (mixed debris). — Large subangular fossil fragments (crinoid, bryozoa, algal?) closely packed and surrounded by calcite cement. Mississippian Mission Canyon fm., Williams Co., North Dakota. Photograph of an etched sample, ×4.

J. Same as I, but a thin section photograph instead of an etched sample, ×8.

K. Crinoidal limestone. Composed of crinoid columnals surrounded by clear calcite cement. Mississippian Madison group, Uintah Co., Utah, ×8.

L. Foraminiferal (Orbitolina) limestone. Composed dominantly of Orbitolinas, skeletal fragments, and detrital grains tightly cemented by clear calcite. Cretaceous Apon, Venezuela, ×8.

M. Coralline algae limestone. Large coralline algae remains scattered in a matrix of fossil debris and micrite. Miocene age, Philippines, ×8.

N. Limestone with algal segments. Contains algal segments up to 5.0 mm diameter, algal spheres, shell fragments, and lumps embedded in a matrix of cement and micrite. Mississippian Charles, Big Horn Co., Wyoming, ×8.

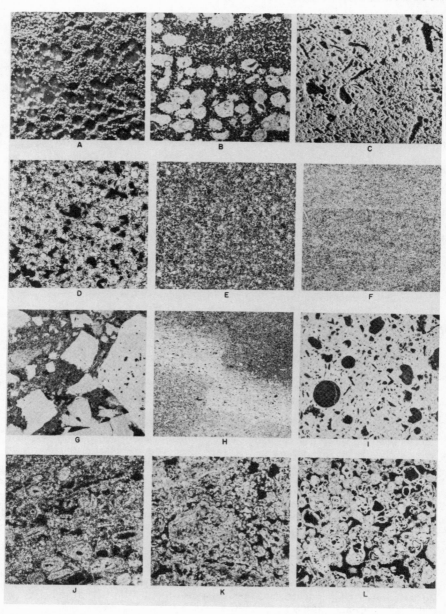

Plate 2-2   Dolomite rock types.

Explanation Plate 2-2. Dolomite rock types[28]

A. Calcareous dolomite (oölitic). Composed of coarse-grained oölites and very fine-grained dolomite crystals with up to 3 percent micritic limestone and calcite cement. Etched section. Mississippian Mission Canyon, Park Co., Wyoming, ×4.

B. Same as A, but a thin section photograph instead of an etched sample. Photo, ×8.

C. Calcareous dolomite (calcitic skeletal grains). Contains skeletal fragments, irregular pellets, molds of fossil debris, and coarsely micrograined dolomite rhombs. Etched sample. Mississippian Charles, Big Horn Co., Wyoming, ×4.

D. Medium-grained dolomite with rhombic texture. Porous network of medium-grained euhedral to subhedral dolomite rhombs. Jurassic Arab, Saudi Arabia, ×8.

E. Very fine-grained dolomite with mosaic texture. Tightly packed mosaic of dolomite crystals. Ordovician Ellenburger fm., Travis Co., Texas, ×8.

F. Finely micrograined dolomite. Texturally uniform, relatively pure dolomite with average crystal size of 0.03 mm. Mississippian Charles, Park Co., Wyoming, ×8.

G. Dolomite breccia. Angular to subangular dolomite fragments in a matrix of micrograined dolomite crystals. Mississippian Mission Canyon, Big Horn Co., Wyoming ×8.

H. Laminated dolomite. Laminae microfaulted in photograph. Laminae contain quartz silt grains. Anhedral interlocking finely micrograined dolomite rhombs. Mississippian Charles fm., Big Horn Co., Wyoming ×8.

I. Dolomite with fossil molds. Vuggy, coarsely micrograined dolomite. Vugs are end product of leaching of fossils. Molds of crinoid and shell debris. Pennsylvanian Hermosa, San Juan Co., Utah, ×8.

J. Dolomitized oölitic limestone. Very fine-grained dolomite with relict coarse-grained oölites, and very fine-grained pellets. Mississippian Lodgepole, Park Co., Wyoming, ×8.

K. Dolomitized algal-lump limestone. Vuggy porosity. Permian Capitan, Winkler Co., Texas, ×8.

L. Dolomitized skeletal-oölitic limestone. Porous dolomite with relict coarse-grained oölites and relict medium-grained Foraminifera. Pennsylvanian Hermosa, San Juan Co., Utah, ×8.

There are several forms of carbonate porosity.[25,26]

1. Intergranular – the void space between grains. It varies depending on particle shape, sorting and alteration. It is the one singularly most important feature for the storage of oil.

2. Intragranular – the space created by the skeletal framework left intact by particular organisms. Badly affected by diagenesis.

3. Secondary void – a rather rare type of porosity where druzy calcite fills the pore spaces after which the original rock grains are leached out.

4. Channel – porosity in the sense that channels are long irregular vugs which vary greatly in dimension.

5. Intercrystalline – where crystals are wide relatively and connected at the apices as in some dolomites.

Studies of the Arab-D formation in the Persian Gulf by R. W. Powers and others[27] points up the complexity of carbonate rock authigenesis. He claims that the Arab-D has been altered at least five and possibly six times in the approximate following order:

1. Druzy calcite coating added to calcarenite particles.
2. Recrystallization of rock to mosaic calcite.
3. Cementation by clear calcite.
4. Growth of anhydrite crystals.
5. Dolomitization.
6. Silicification (chert)

2.2 Carbonate Energy Index Classification

When enough data are available to develop the sedimentation history of a lithologic column, the forces which contributed to the texture and composition of that column can be classified, thus shedding light on the rock's environment and oil-bearing potential.

A detailed energy index for limestone interpretation and classification was developed by Plumley, Risley, Graves and Kaley.[28] They presume limestones are mainly biogenic and originate within the depositional basin. The energy index they have compiled relies on the study of primary rock features discerned even among secondary crystallization and mineralogical changes. The key to classifying a sediment lies in the biotic and textural makeup of the rock.

In shortened form their several definitions required for energy index interpretation are:

1. Wave base – the water depth below which surface waves do not move sediment.

TABLE 2-1 ENERGY INDEX CLASSIFICATION OF LIMESTONES

Limestone type according to energy index	Limestone sub-type (Plates 7–11)	Mineralogy	Size	Sorting	Roundness	Fossil abundance and complexity	Characteristic fossils, fossil associations, fossil preservation
Quiet I Deposition in quiet water	I1	Calcite Clay (15 to 50%) Detrital quartz (<5%)	Microcrystalline carbonate (<0.06mm) or any size fossil fragments in a microcrystalline carbonate matrix (matrix <50%)	Matrix—good Fossils—poor	Original fossil shapes; angular fragments if broken.	Barren to moderately fossiliferous, simple assemblages	Crinoids; echinoids; bryozoans (fragile branching types); solitary caroals; ostracodes; thin-shelled brachiopods; pelecypods, and gastropods; Foraminifer; sponge spicules; tubular, encrusting, and sediment-binding algae; fecal pellets of bottom scavengers. Common fossil associations are crinoid-bryozoa assemblages bivalve shell assemblages. Foraminifera assemblages (predominantly planktonic).
	I2	Calcite (predomenant) Clay (<15%) Detrital quartz (<5%)		Matrix—good Fossils—moderate to good		Moderately to abundantly fossiliferous, simple assemblages (Coquinoid limestone)	
	I3		Any size fossil fragments in microcrystalline matrix (matrix <50%)				Many fossils are whole and unbroken and are not mechanically abraded. Any fragmentation of fossil material probably is due to disarticulation upon death, to predatory (boring, opening, and breaking) activity and scavenger activity or to solution.

TABLE 2-1 *(Continued)*

Limestone type according to energy index	Limestone sub-type (Plates 7-11)	Mineralogy	Size	Sorting	Roundness	Fossil abundance and complexity	Characteristic fossils, fossil associations, fossil preservation
Intermittently agitated II Deposition alternately in agitated water and in quiet water	II1		Microcrystalline matrix (>50%) Micrograined to medium-grained clastic carbonate and terrigenous material	Matrix—good Clastic material—poor to good	Clastic carbonate material subangular to rounded. Roundness of terrigenous clastics is principally a function of size. Oolites may be present	Barren to moderately fossiliferous, moderately simple assemblages	Characteristic fossils and fossil associations are similar to Type I limestones. Fossil materials are more fragmental than those in Type I limestones and also may be more or less rounded by wave action. Scattered fragments of fossils from rougher water environments may be present.
	II2	Calcite (predominant. Clay (<25%) Detrital quartz (<50%)	Microcrystalline matrix (>50%) Coarse to very coarse-grained clastic carbonate and terrigenous material				
	II3		Interbedded microcrystalline carbonate and any size clastic. Microscale rhythmic bedding	Sorting good within individual lamina		Barren to moderately fossiliferous, moderately complex assemblages	
Slightly agitated III Deposition in slightly agitated water	III1		Micrograined clastic carbonate (<0.06 mm) predominates	Matrix—good Clastic material—moderate to good	Clastic material subrounded to well rounded Fine-grained oolites may be present	Barren to sparsely fossiliferous, simple assemblages	Echinoderm, bryozoan, and bivalve shell debris; Foraminifera; encrusting algae. Common fossil associations are Foraminifera-abraded bivalve shell fragment assemblages. Fossil materials comminuted from larger fossil structures are well abraded by wave and current action.
	III2	Calcite (predominant) Detrital quartz (up to 50%)	Very-fine-grained clastic carbonate (0.06 to 0.125 mm) predominates	Matrix—poor Clastic material—moderate to good		Barren to moderately fossiliferous, simple assemblages	
	III3		Fine-grained clastic carbonate (0.125 to 0.25 mm) predominates			Barren to abundantly fossiliferous, simple to moderately complex assemblages	

Environment	Code	Clastic material	Mineralogy	Matrix	Rounding	Fossil content	Fossil assemblage
Moderately agitated IV Deposition in moderately agitated water	IV1	Medium-grained clastic carbonate (0.25 to 0.5mm) predominates		Matrix – poor Clastic material – moderate to good	Clastic material subrounded to well rounded. Oolites may be present	Moderately to abundantly fossiliferous, simple to moderately complex assemblages	Crinoids, echinoids, bryozoans, brachiopod and pelecypod shell fragments, colonial coral fragments, stromatoporoid fragments (Silurian and Devonian predominantly) tubular algal fragments, colonial algal fragments (rare) encrusting algae. Common fossil associations are similar to association of Types I, II, and III or they are mixtures of these associations. Fossil materials are generally broken and abraded.
	IV2	Coarse-grained clastic carbonate (0.5 to 1.0 mm) predominates	Calcite (predominant) Detrital quartz (up to 50%)			Moderately to abundantly fossiliferous, moderately complex to complex assemblages	
	IV3	Very coarse-grained clastic carbonate (1.1 to 2.0 mm) predominates					
Strongly agitated V Deposition and growth in strongly agitated water	V1	Gravel-size clastic carbonate (rock fragments and fossil material 2.0 mm) predominates		Matrix – poor Clastic material – poor to moderate	Clastic material subrounded to well rounded. Pisolites may be present	Sparsely to moderately fossiliferous, complex assemblages	Crinoids; echinoids; encrusting bryozoans; thick-shelled brachiopods pelecypods; colonial coral fragments; stromatoporoid fragments (Silurian and Devonian predominantly); colonial algal fragments; rudistid fragments (Cretaceous predominantly). Fossil associations are similar to Type IV. Fossil materials are generally broken and abraded.
	V2	Gravel-size conglomeratic or brecciated carbonate (2.0 mm) Tectonic breccias excluded	Calcite (predominant) Clay (<5%) Detrital quartz (<25%)	Matrix – poor Clastic material poor	Clastic material angular to well rounded	Barren to sparsely fossiliferous, complex assemblages	

TABLE 2-1 *(Continued)*

Limestone type according to energy index	Limestone sub-type (Plates 7–11)	Mineralogy	Size	Sorting	Roundness	Fossil abundance and complexity	Characteristic fossils. fossil associations. fossil preservation
	V3	Calcite	Not applicable	Not applicable	Not applicable	Abundantly fossiliferous, simple assemblages (fossil colonial growth in place).	Colonial corals, stromatoporoids, colonial algae (principally the Rhodophyth or red algae and some genera of the Cyanophyta or blue-green algae).

From AAPG Memoir I *Classification of Carbonate Rocks*: W. J. Plumley, G. A. Risley, R. W. Graves, Jr., and M. E. Kaley. Energy Index for Limestone Interpretation and Classification, p. 88–89. Reproduced with permission from the American Association of Petroleum Geologists.

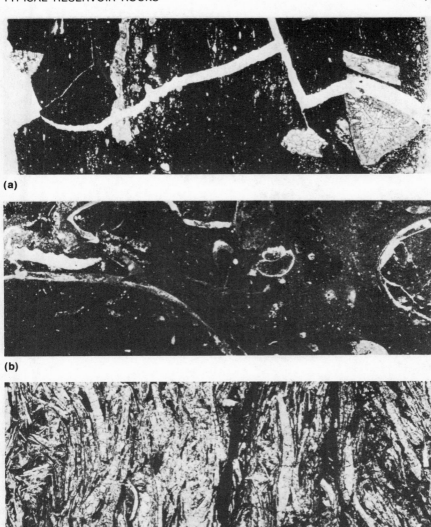

(a)

(b)

(c)

Plate 2-3 Type 1 limestone.[28] (a) Type I$_1$ microcrystalline limestone with large bryozoan fragments and about 45% clay. Pennsylvanian, Strawn Series; Sotex Brown 2–5, depth 7,545 ft; Scurry County, Tx.; (b) Type I$_2$, slightly fossiliferous microcrystalline limestone. Devonian, Woodbend formation; Socony Vegreville 1, depth 3,069 ft; Alberta, Canada.; (c) Type I$_3$, coquinoid limestone, bivalve shells in a microcrystalline carbonate matrix. Cretaceous, Cogollo formation; Rexco Zulia 26D-2, depth 11,670 ft; Venezuela.

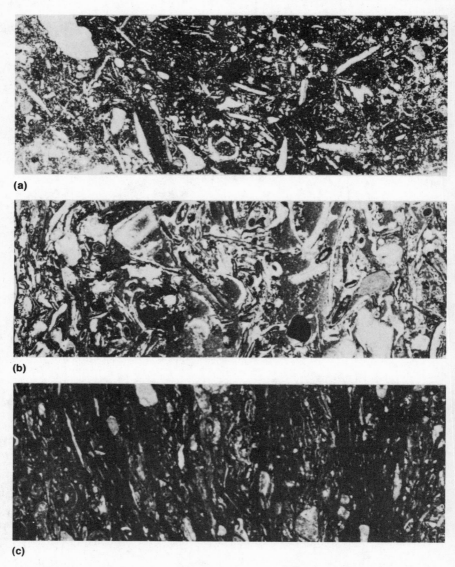

(a)

(b)

(c)

Plate 2–4 Type II limestones.[28] (a) Type II_1, fine to medium-grained fossil debris in microcrystalline carbonate matrix. Cretaceous, Cogollo formation; Texas Raban 10, depth 8,260 ft; Venezuela. (b) Type II_2, coarse to very coarse-grained rock fragments and fossils in microcrystalline carbonate matrix. Devonian, Woodbend formation; Socony Vegreville 1, depth 3,041 ft; Alberta, Canada; (c) Type II_3, interlaminated microcrystalline carbonate and fossil debris. Mississippian; Cal Standard West Daly 8–29, depth 2,474 ft; Manitoba, Canada.

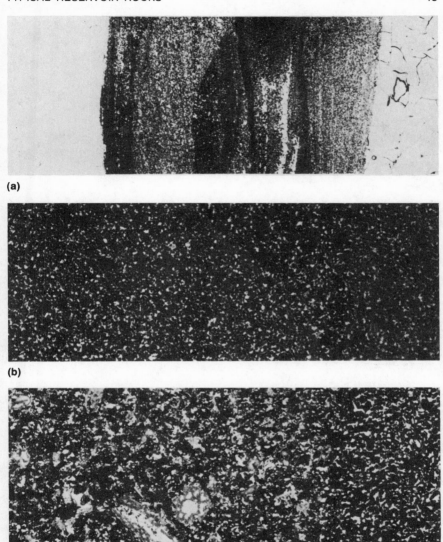

(a)

(b)

(c)

Plate 2-5 Type III limestones.[28] (a) Type III$_1$, microclastic limestone with well-developed cross bedding. Mississippian; Canadian Superior Cruikshank 14–4, depth 2,712 ft; Manitoba, Canada; (b) Type III$_2$, very fine-grained clastic limestone. Jurassic, Smackover formation; Stanolind Bodcaw 1, depth 11,050–51 ft; Lafayette County, Arkansas; (c) Type III$_3$, fine-grained clastic limestone, Devonian, Woodbend formation; Socony Vegreville 1, depth 3,871 ft; Alberta, Canada.

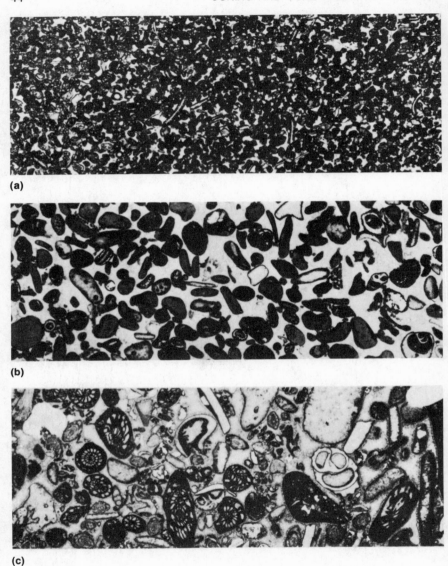

(a)

(b)

(c)

Plate 2–6 Type IV limestones.[28] (a) Type IV₁, medium-grained clastic (algal oölite) limestone, Pennsylvania, Virgil Series; survace sample; Sacramento Mountains, New Mexico; (b) Type IV₂, coarse-grained clastic (rock and fossil fragments) limestone crystalline calcite matrix. Cretaceous, Cogollo formation; Texas Raban 10, depth 7, 867 ft; Venezuela; (c) Type IV₃, very coarse-grained clastic limestone with crystalline calcite matrix. Pennsylvanian, Canyon Series; Sotex Brown 6–6, depth 6,770 ft; Scurry County, Tx.

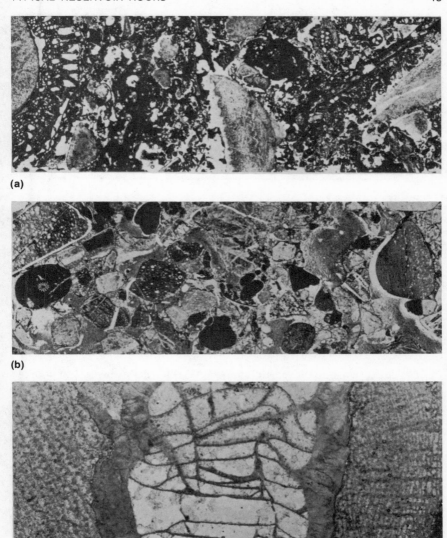

Plate 2–7 Type V limestone.[28] (a) Type V_1, gravel-size clastic limestone. Pennsylvanian Canyon Series; Sotex Brown 3–1 depth 6,677 ft; Scurry County, Tx; (b) Type V_2, conglomeratic limestone. Pleistocene, Ryukyu formation; surface sample; Ie Shima, Ryukyu Islands; (c) Type V_3, stromatoporoid and colonial coral limestone. Devonian, Woodbend formation; Socony Vegreville 1, depth 3,818 ft; Alberta, Canada.

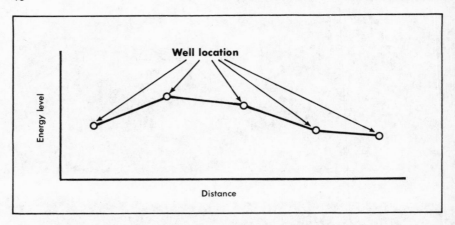

Figure 2–2 Diagram to show energy levels from samples with respect to position in the lithologic column. Each energy level can be subdivided into the more detailed units described in Table 2–1. After diagram by Plumley, Risley, Graves and Kalye, AAPG Memoir 1, Classification of Carbonate Rocks, p. 101.

2. Energy level—kinetic energy available at the depositional interface and a few feet above.

3. Depositional interface—the zone between the water and the bottom where sediment is deposited relative to the energy level at the interface.

4. Clastic carbonate particles—grains that have been transported mechanically by current or wave action.

5. Microcrystalline carbonate—particles less than 0.06 mm diameter that cannot be called clastic.

6. Micrograined carbonate—particles between 0.0039 and 0.06 mm diameter that can be interpreted as clastic grains (silt-sized).

7. Matrix—material in which particles are embedded.

8. Nonskeletal grains—silt to sand-sized particles whose genesis is unknown.

Table 2–1 describes in great detail the five energy levels created by currents and waves. Plates 2–3 to 2–7 show examples of the five energy zones. According to Plumley et al.,[31] "carbonate sediments respond to the same energy levels as do the terrigenous clastic sediments. The coarser-grained clastic carbonates are found in the higher-energy environments at the beach zone and in shoaling waters created by carbonate banks and reefs. The low-energy carbonates are found in shallow back-reef or lagoonal areas or in deep water seaward of carbonate bank or reef buildups."

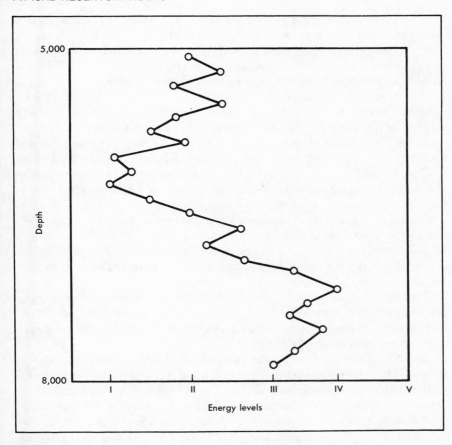

Figure 2-3 Schematically drawn energy levels of a particular formation as determined from a series of wells. Particularly useful in displaying drilling development of a reef. After diagram by Plumley, Risley, Graves and Kaley, AAPG, Memoir 1, Classification of Carbonate Rocks, p. 105.

The energy index can be drawn in relation to the lithologic column as in Fig. 2–2 or as in a cross-sectional diagram of Fig. 2–3.

2.3 Sandstones

Sandstones are made up of the detritus of crystalline plutonic rocks, generally of terrigenous origin. Before becoming sand or sandstone as seen in a particular core, they may have been metamorphosed or have been an earlier generation sandstone formation.

The "supracrustal" rocks (low-grade metamorphics, volcanics and sediments) contribute rock particles which loggers call dirty minerals. The plutonic variety of rock contributes quartz and feldspar. Feldspar is thus a useful source rock index in that the relative amounts of feldspar to dirty minerals indicate the type of source rock.

Feldspar and the dirty minerals (mica, chlorite, glauconite sericite (altered muscovite), kaolinite) all have a hardness of 2.5 to 3 where quartz has a hardness of 7. Where the dirty minerals are absent and quartz predominates, one invariably finds less angular and more subrounded quartz grains indicating well worked or reworked sedimentary material.

The terms "sands" and "sandstones" are used concurrently although they should be kept separate in the strict sense of their meanings. Sands are unconsolidated and made up of angular to subrounded quartz grains with little or no cement. Well-rounded quartz sands are called loess and are deposits from wind erosion.

Sandstones are made up of one or several types of minerals; quartz is almost always the dominant mineral and the grains are angular to subrounded, cemented either partially or completely.

The color of sandstone is usually dictated by the color of cementing material which in turn indicates the depositional environment of the period. When sandstones are fractured, they break apart along the surfaces of the grain material, not through it.

In an ordinary sandstone the possible void space between grains is about 32% of the whole mass. The voids may be filled wholly or partially with precipitated cement, fine silts or clay.[32] Fig. 2–4 shows a classification of sands and sandstones which is readily adaptable to wellsite work.

An important fact to remember about sandstones and sands is that their interpretation pivots around the quartz content and the type and quantity of matrix cementing material. The more quartz there is in a sandstone, the more mature is the formation.

Lithic sands (Graywacke): These occur in relatively thin beds with no lineations but often with graded bedding. This, it is appropriately assumed, indicates short-lived catastrophic events. These sands are associated with geosyncline belts and tectonically unstable areas. The sediments are marine in nature and deposited by submarine turpidity flows, in the reducing zone of the sea.

A proper definition is difficult to develop because of complex composition and similarity to other sands into which it grades so easily. Texture is perhaps the best way to recognize and describe it. Dark grey and even black-looking when fractured because of the detrital quartz, the rock contains sharp, angular, spalled-off quartz slivers with angular feldspar grains

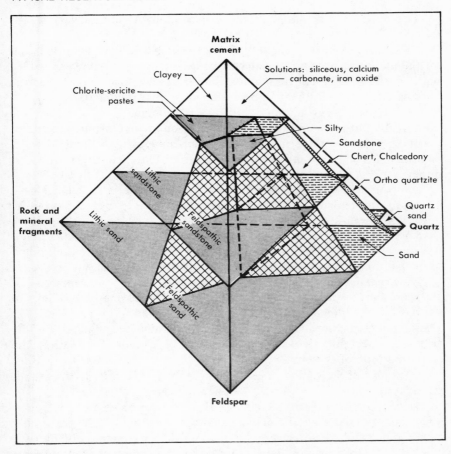

Figure 2-4 Model of the province of sand and sandstone.

and dark mineral particles. These constituents are set in a matrix made up of a microcrystalline aggregate of quartz, feldspar, chlorite, and sericite which at times may be replaced or be substituted by a carbonate cement.

Lithic sandstone (Graywacke): The simple definition says there are more rock and mineral fragments than feldspar particles. The voids between the sedimentary fragments are filled with clay and/or mineral cements. It commonly has calcium carbonate cement.

The grains are well rounded to sub-rounded, normally, and porosity is rather high. The beds may be crossbedded or finely stratified. Yet, because of the amount of feldspar content, the sediments must still be considered immature.

The sediments are thought to be deposited by normal subaqueous currents.[33]

Feldspathic sands (Arkose): These occur as thin residual beds immediately above granitic rock or as wedge-shaped deposits interbedded with granite-bearing conglomerates. The rocks are formed by terrestrial current deposition and may appear almost as the original granite.

The sediments form in an oxidizing environment.

Feldspathic sandstone (Arkose): This is one where orthoclase and/or plagioclase feldspar is a major constituent mineral with voids filled by siliceous and/or carbonate cementing material. The rock is terrigenous in origin.

Most oil-field sandstones are devoid of feldspar because, being marine in nature, the feldspars are eroded away by the time they could be deposited in an oil-forming marine environment.

Sandstones with a reasonable amount of feldspar are found in granite washes where some oil is found. Otherwise the feldspars are only noted in the gas sands which exhibit coastal swampy conditions.

Sand, sandstone: These two simple words describe the majority of clastic reservoir material examined by the geological engineer. They are the basic descriptions upon which a composite picture of a rock is built. The complete picture includes the size of the grains, the angularity of the grains, any secondary coatings on the grains, the color of the grains, the color and type of cementing material etc.

The range of the sand and sandstone is shown by Fig. 2–4. Most sandstones and sands contain 60–90% quartz grains.

Orthoquartzite, Chert, Quartz Sand. . . . These rocks are made up of 90% or more quartz and/or silicious material. Orthoquartzites invariably have the voids filled with cementing material. They are normally found as blanket sandstones which have been well sorted of associated minerals and exhibit by their grain sphericity and cleanliness, an extremely mature formation.

Orthoquartzites commonly grade into limestones and dolomites and because of their lack of associated minerals, represent highly concentrated decomposition at the time of deposition and/or several cycles of quartz-enriching sedimentation.[34]

In Northern California, ancient sandbars are now orthoquartzite "shells" which are 0.5 to 5 ft thick and are found at random when drilling.

Most shells exhibit a form of hydrous metamorphism where anhedral quartz grains lock together from growth of secondary quartz.

Chert, technically a cryptocrystalline granular quartz is formed from siliceous solutions and plays a part as a "sediment" when it becomes

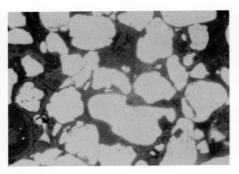

Plate 2-8

Plate 2-10

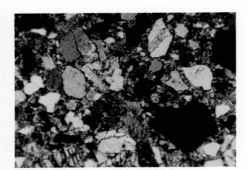

Plate 2-9

Plate 2-8. Hematite-cemented sandstone, Hythe Beds, Godstone, Surrey. Plain polarized light, X10. Grains are subrounded to rounded. Notice that all the grains seem to "float" in the hematite matrix. (Courtesy of Dr. J. F. Potter, F.G.S., Banta © 1973)

Plate 2-9. Calcite-cemented sandstone, Largs, Ayrshire. Crossed nicols. X10. Subangular to subrounded, fine to coarse grains. Some grains show grain corrosion. Some quartz appears grey, white, or black depending on the crystal axis orientation relative to the light. Notice the veinlet of calcite across one quartz grain and the larger vein of calcite diagonally across the photo. Some feldspar grains, including plagioclase, are present. (Courtesy of Dr. J. F. Potter, F.G.S., Banta © 1973)

Plate 2-10. Chert, Hythe Beds, Sundridge, Sevenoaks, Kent. Crossed nicols. X20. Cryptocrystalline fibrous, radially orientated chalcedony in which the black holes indicate areas in which the crystal axes are orientated normal to the photograph. Calcite evident in one corner of the picture. Dark line is a sponge spicule and the prominent fossil is a bryozoan preserved in calcite. (Courtesy of Dr. J. F. Potter, F.G.S., Banta © 1973)

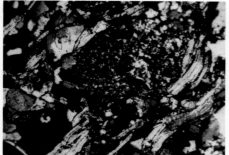

Plate 2-11

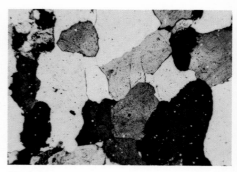

Plate 2-13

Plate 2-12

Plate 2-14

Plate 2-11. Micaceous sandstone. (Grey-wacke), Downtonian, Llandeilo, Wales. Crossed nicols. X10. Angular quartz grains. The orientation of the muscovite indicates the bedding planes of the rock. A large siltstone pebble is visible near the center of the photograph. Chlorite and clay dominate the matrix material. (Courtesy of Dr. J. F. Potter, F.G.S., Banta © 1973)

Plate 2-12. Fine-grained feldspathic sandstone, Basal Caradoc, Tarn Hows, Lake District. Crossed nicols, X10. Some of the feldspar grains are mottled and very corroded. Very fine grains with silica make up matrix material. (Courtesy of Dr. J. F. Potter, F.G.S., Banta © 1973)

Plate 2.13. Compacted and silicified sandstone, Hartshill quartzite, Hartshill, Leicestershire. Crossed nicols, X10. Subangular to subrounded quartz grains compacted together to the extent that grain boundaries are altered. Slight carbonate material evident interstitially. (Courtesy of Dr. J. F. Potter, F.G.S., Banta © 1973)

Plate 2-14. Sandstone, Mauchline, Ayrshire. Crossed nicols, X10. Subrounded to rounded quartz grains, showing secondary silica enlargement. Black areas are also grains not favorably exposed due to crystal axes orientation relative to the light. (Courtesy of Dr. J. F. Potter, F.G.S., Banta © 1973)

fractured and capable of holding oil as it does at Ventura, Calif. There are various opinions about the formation of chert because of the abundant diatomaceous or radiolarian material found in it. Bedded chert probably develops from siliceous gels on a sedimentary floor into which organisms have come to rest. Chalcedony is very much like chert except it is fibrous rather than granular. The two are often mistaken for each other and the mineral names are often used interchangably.

Quartz sand is made of pure quartz grains, round frosted and with no cementing material. In parts of the Persian Gulf the Burgan Sand is an example of a quartz sand. This sand is the Cretaceous sand so prolific with 31.0° API gravity oil in Kuwait that the country has become one of the wealthiest per capita in the world. North of Quatar a core was taken of this same sand and found to contain 5° API gravity oil. The core was held together by the oil and when it was cleaned away, the quartz grains rolled about like ball bearings — it wasn't noticed before the cleaning process that there was no cementing material. Such loess-like sands are terrestrial in origin, at least in the last cycle of their formation. To attain such sphericity and cleanliness requires the grains to be reworked several times.

Matrix cement. The sources of cementing material and how and when grains are cemented are unresolved questions. In Paleozoic times about 75% of the cementing material was siliceous in origin. Since Mesozoic times silica and carbonate cements share the cement distribution.

Plates 2–8 to 2–14 are sandstone thin sections printed in color to emphasize the grain orientation and make mineral differentiation easier.

3

Gas, Oil,
Water Mechanics

The interaction of water, oil, and gas in a reservoir is critical because production can be hindered or stopped altogether by inadvertently altering the balance of these constituents. It serves our purpose in this chapter to become familiar with the different types of water and oil which are found in reservoirs.

The forces which hold the constituents in balance and force migration under certain circumstances are discussed with regard to their effect on porosity, permeability, and fluid saturation. Reservoir energy must overcome two forces: the force holding the oil or gas in the pore system, and the viscous resistance of the gas and oil to movement.

3.1 Phase Relations

Since hydrocarbon reservoirs are a complex multicomponent system, an understanding of phase relationships between oil, gas, and water is beneficial, if not necessary. The terminology is unique to the wellsite geologist, so some definitions are needed to improve understanding.

1. Bubble point: A condition that exists when a liquid begins to form a gas.

2. Critical point: A condition of pressure and temperature beyond which a liquid must become a gas.

3. Dew point: Where a liquid remains in equilibrium with a gas.

4. Bubble-point curve: A schematic line along which a gas first comes out of a solution.

5. Dew-point curve: A schematic line along which a liquid first begins to condense from a gas.

6. Cricondentherm: The point of maximum temperature at which liquid and gas phases can coexist.[35]

Fig. 3–1 is self-explanatory.

Fig. 3–2 shows a typical pressure-temperature relationship as applied to three wells with different pressure-temperature conditions. Note that in well B which represents a retrograde condensate pool, a drop in reservoir pressure will put fluid conditions along the dew-point curve. Liquid formed along this curve or below it, is nonrecoverable because it becomes

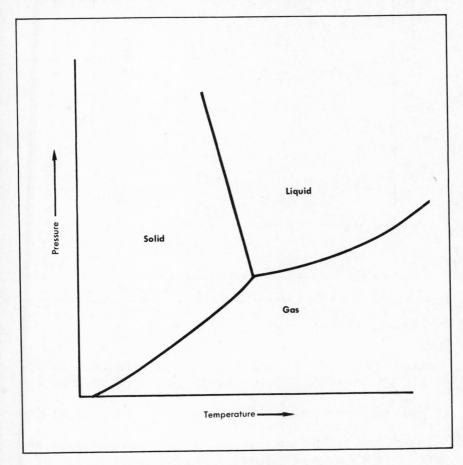

Figure 3–1 Water system phase diagram, valid for low pressures only.

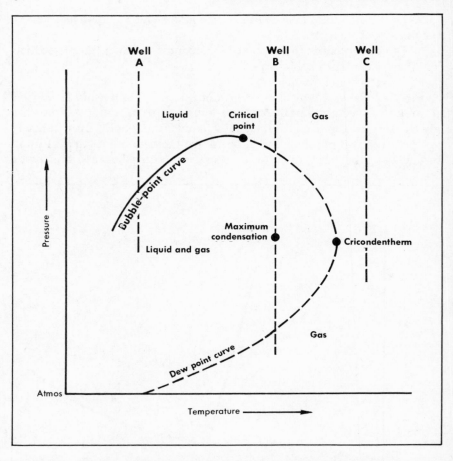

Figure 3–2 Pressure-temperature diagram for a hypothetical two-component system of crude oil and natural gas at constant composition. A — oil well; B — retrograde condensate pool; C — gas well. From Levorsen, A. I., Geology of Petroleum, second edition.

bound as a film to the reservoir rock through molecular adsorption. The gas which was dissolved in the oil at high pressure, passes to the well and is produced.

Well "C" remains a gas well because pressure changes don't affect it.

In well "A," as oil is produced, the pressure drops until the bubble-point is crossed when gas accompanies oil in production.

As a well is produced, the phase relationships between the fluid and gas phases are continuously upset when one or more of the phases is produced disproportionately to the other(s).

3.2 Capillarity

Capillary pressure has two important effects in oil and gas reservoirs. First, it controls the distribution of fluids in an untapped reservoir. Secondly, it is the mechanism whereby oil and gas move through reservoir pore spaces until they are confronted with a barrier.

Capillary pressure may be defined as the differential pressure that exists between two fluids (two liquids, or a liquid and gas) as a result of the interface curvature that separates them.

The difference in pressure equals the surface tension of the interface. The lower pressure occurs on the convex side of the two phases of immiscible fluids (oil and water). This side is usually water which, in most wells, is the wetting phase. This means that it coats the grains of rock over which the oil or liquid gas must flow.

Conversely, oil or gas is termed the nonwetting phase since it does not initially come in direct contact with the rock grains. Vugs and fractured sediments hold little capillary water when they contain oil.

If oil, forced into a water-wet core or a rock fragment saturated with water, drives out and replaces the water, the relationship between saturation and capillary pressure follows the pattern of Fig. 3–3. In curve 1 the pressure needed to force the oil to enter the rock — displacement or entry pressure — must rise from A to B. Curve 1 shows the rock to be a clean, uniform-grained, and well sorted, with a permeability of 100–200 md.

The curve from B to E, being nearly flat shows that it takes very little additional pressure to increase the oil saturation (decrease the water saturation) greatly. An increase in pressure from E to F, however, adds only about 10% oil and, from there on to C, additional pressure does not reduce the water content appreciably.

The water content of DC is called the residual water (the irreducible water saturation). Most of it is confined to the pendular areas. The changing position of the oil-water interface within a single capillary pore as the saturation changes is shown in the inset, where the letters correspond to the same letters on the graph (W = water).

Curve 2 is characteristic of a limestone or dolomite with 15 to 25-md permeability.

Curve 3 is indicative of a gradation in size of grains and pores and a variable permeability due to the presence of clay and matrix material; a steady increase in pressure is needed to force the oil into the rock, and

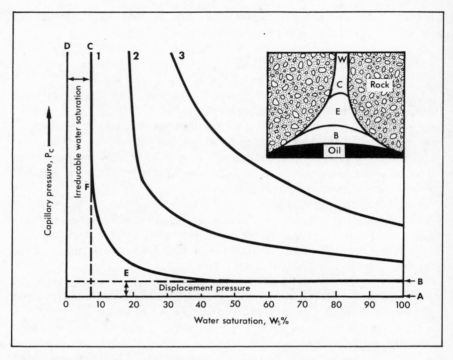

Figure 3–3 Examples of capillary pressure curves.

the residual water content is high because of the larger amounts of water held in the finer pores.[37]

3.3 Wettability

Wettability, sometimes called energy of adhesion, is important for the consideration of the oil-water interface. The interface contact with the rock grain surface is measured through the water phase. If the angle is acute through the water phase, it is the wetting agent since its energy of adhesion is greater. If the angle Θ is greater than 90° the rock is considered nonwet.

The contact angle Θ will vary depending on whether a liquid is advancing or retreating over a given rock surface. Natural gas is nonwetting, oil is wetting when compared with gas, and oil is nonwetting when compared with formation water.[38]

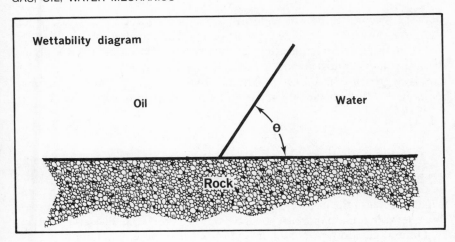

Depending on the type of rock in the reservoir, the wetting capabilities will differ. Most reservoirs have water as the wetting phase.

With existing theories of oil migration and accumulation, the oil must displace the water which occupies the capillary openings. The minimum pressure which will force oil into the openings is called the displacement pressure. The smaller the pore, the greater the pressure required to force a globule of oil into its space. Thus, the wetting properties of a rock are very important in their effect on relative permeability.

The wetting fluid will preferentially cover the entire solid surface of the reservoir rock and will be held in the smaller pore spaces of the rock because of the action of capillarity. The nonwetting phase (oil) of the two phases tends to be repelled from the rock's surface. So, at small saturation levels, the nonwetting phase will tend to collect in larger pore spaces.

3.4 Oil-Water Transition Zone

The transition zone from water to oil varies in thickness depending on the capillary pressure. Since oil has a lower specific gravity than water, it displaces the water already in the rock as it migrates from its source. This displacement of the water downwards continues until the water saturation in the section of rock being invaded by the oil is reduced to a point where it will no longer flow and the oil's displacement pressure is equalized.

It can thus be expected that the oil-water interface angle will not be so acute in the transition zone. Oil exists there in the largest pores as globules whereas farther up the formation, oil saturation is greater, which increases oil-water interface curvatures and capillary pressures.

Irreducible water saturation is always found in oil reservoirs above the water table and transition zone. Low-permeability rocks have high capillary pressure and long transitional zones whereas high-permeability rocks have low capillary pressure and short transition zones.

3.5 Buoyancy

The difference in density of oil and water results in the oil being trapped above the water. As the oil invades the water-wet trap at the termination of a migratory period, its buoyancy increases with each additional droplet of the nonwetting fluid. This causes the capillary pressure of the reservoir rock to increase at all elevations above the oil-water contact.

Oil and gas do not invade the caprock since there is not enough pressure in the buoyant zone to allow it to pass into the caprock pore spaces. Only if by rare chance the reservoir was sealed and water could no longer be displaced downwards and the pressure was increased by tectonic action or further accumulation of oil, could oil and gas acquire the capillary pressure necessary to invade the impermeable caprock.

Once sedimentary compaction has slowed or ceased and the sediments are lithologic units, the movement of formation water slows dramatically. One exception, seen personally, is an area in the Persian Gulf where, drilling at 11,000 ft, one experienced the disappearance of a complete section of the mud column every time the trip was made for a bit change. Black brackish water replaced the mud. This particular phenomenon continued to occur even after drilling below the migrating fluids.

Faulting, folding and grain cementation become inhibiting factors to migration, forming constrictions which develop local pressures. However, the physical law governing buoyancy states that a fluid or solid which is immersed in a fluid is buoyed by a force equal to the weight of the fluid it displaces. Thus, when there is enough concentration of oil and gas, it begins to move upwards as the buoyancy force overcomes capillary pressure resistance to the entry of oil and gas into the water-saturated pore spaces.

Buoyancy plays an important role in the several theories of migration. Some experts say that water is required to move oil and gas. Others claim that oil and gas can migrate alone without the wate phase. Oil at 70°F and at 500 psi is, when saturated with gas, half as viscous as the same oil-gas combination at surface conditions.

Figs. 3-4 to 3-7 show changes in water saturation with height above

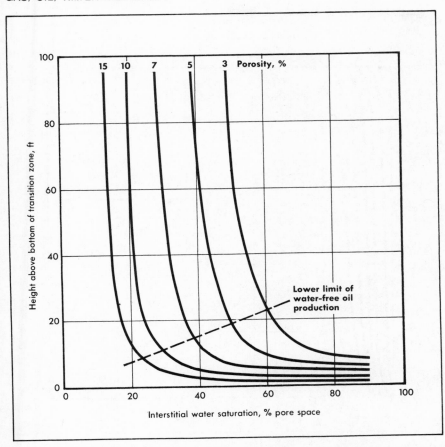

Figure 3-4 Water-distribution curves for Pennsylvanian dolomite, for porosities of 3–15%. The dashed line represents the imaginary plane in the formation above which oil is produced without accompanying interstitial water. From The Fundamentals of Core Analysis, Core Laboratories, Inc.

the transition zone and demonstrate the influence of buoyancy for dolomites in a wide range of porosities and permeabilities.

3.6   Classification of Oilfield Waters

There are several types of water found associated with oil and gas wells. They are highly suspect in that it cannot be supposed that they existed at the time of deposition as they now chemically exist.

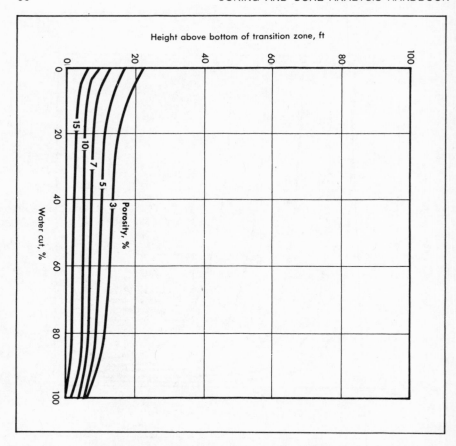

Figure 3–5 Data from the same Pennsylvanian dolomite as in Fig. 3–4. It represents the connate and other waters mixed with the oil above the oil/water transition line, showing the effect of porosity in the buoyancy process of oil-water separation. From The Fundamentals of Core Analysis, Core Laboratories, Inc.

The assumption is that sedimentary basin waters today are no different than they were eons ago when the oil was formed and trapped. What has happened to many brines which are produced with oil is that they have undergone various types of chemical change depending on the composition of the source rocks and reservoir rocks, and the heat and pressure associated with the depth of burial.

Under specific conditions, waters high in chemical content may alter the mineral content of a rock. In the exchange of ions between rock min-

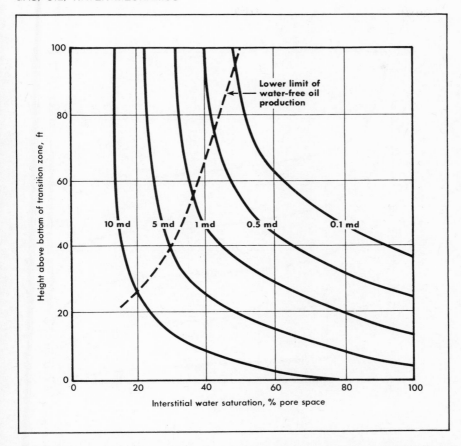

Figure 3–6 Water-distribution curves for San Andres dolomite, for permeabilities of 0.1–10.0 md. The dashed line represents the imaginary plane in the formation above which oil is produced without accompanying interstitial water. From The Fundamentals of Core Analysis, Core Laboratories, Inc.

erals and water, pressure increases the power of adsorption and heat decreases it. However, both pressure and heat increase the solubility of salts in water.

There are two basic differences between seawater and oil-field brine. First, sulfate (SO_4^-) is present in seawater but is absent from some oil-field brines. Secondly, the alkaline earths (Ca & Mg) are present in seawater and are absent from certain oil-field brines.

Meteoric waters are those from rain, fresh-water streams, lakes and the like. Connate waters are found in the reservoir rock. They are in the

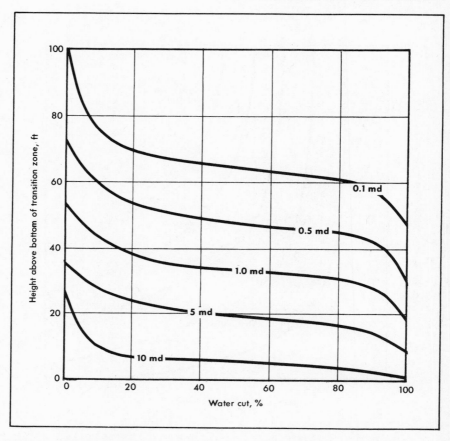

Figure 3-7   Data from the same San Andres dolomite as in Fig. 3–6 showing the effect of permeability in the buoyancy process of oil-water separation. From The Fundamentals of Core Analysis, Core Laboratories, Inc.

pore spaces from the time of sedimentary deposition. This does not take into account that, like oil, the water may have migrated from another source rock because of tectonic action or a change in capillary pressure.

Mixed waters are those with both chloride and sulfate, carbonate-bicarbonate content. These are often found near the surface and immediately below unconformities.

Free Waters are flowing waters sometimes called artesian waters. It is water with a pressure gradient.

Interstitial waters are found with oil and gas and are not displaced by

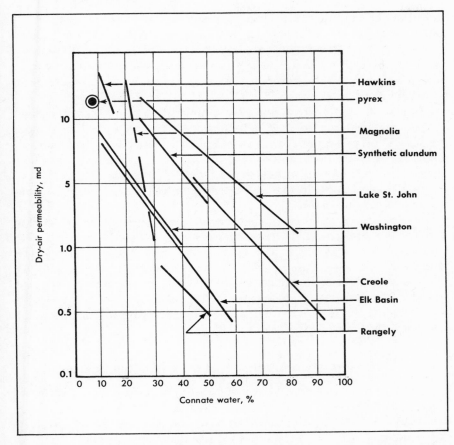

Figure 3-8 From Bruce and Welge, Production Practice and Technology, 1947, p. 170, Fig. 9.

oil and gas when it moves into the upper portions of reservoir traps. In oil-producing zones, anywhere from 10 to 50% of the pore space is occupied by interstitial water. A good example of this phenomenon is the Rangeley Pool in Colorado where almost 50% of the pore space up to 600 ft. above the oil-water contact is taken up by interstitial water.[42]

Interstitial water is an unimportant parameter in fractured reservoir production because the largeness of the fractures override the effect of pore-bound interstitial water capillary pressures. Irregularly sized pore spaces in crystalline, carbonate rocks, which have a high water content, hold the water while the oil moves through the larger pore spaces. This does not apply to sandstone reservoirs because the mean pore-space size in a given piece of sandstone rock is much more uniform and on average, larger than those found in carbonate rocks. Thus oil must pass through the pore spaces with the water. There is a relationship between grain size, interstitial water and permeability. As the permeability decreases, water in the pore space increases.

Interstitial and connate waters are much the same except for slight differences as shown in their definitions. As has been explained under capillarity, oil and water coexist in porous formations in a way dependent upon the pressure between the two liquids. Their coexistence may also depend on which fluid was first in the rock.

The results shown in Fig. 3–8 clearly exhibit the relationship of connate water and permeability.

When water is the wetting agent, it acts as a lubricant for oil passing through the pore space. The water's resistivity can cause misinterpretation of electric logs where the salinity increases. With an increase in salinity there is a decrease in resistivity and oil zones of low conductivity can be missed through miscalculation based on assumed rather than actual saline content of the connate-interstitial water. One can see from Fig. 3–9 that the resistivity increases with a decrease in connate water content as well as a decrease in salinity.

3.7   Oil formation, Characteristics, and Classification

Where oil comes from and how it collects in pools are the two questions both theorists and commercial oilmen have been trying to answer for many years. The province of oil is so complex that all the mystery surrounding it will not be discovered for some time to come and will indeed be an anticlimax for all those engaged in its search.

Inorganic theories of formation have been abandoned by all except a handful of theorists. One must remain aware of inorganic origins such as close proximity of basins to volcanoes and fumaroles which, like the organic origin theory requirement, are a source of hydrogen for oil formation. Organic matter contains between 7 and 10% hydrogen and oil contains 11 to 15% hydrogen.[45] Probably, the likely source of hydrogen is bacteria.

Organic theories of petroleum formation abound. Because of the porphyrins present in most petroleums, a top temperature limit of 200°C. has been established for the creation of oil. The porphyrins also show that anaerobic conditions developed very early in the formation of oil.[46]

Heat and pressure play an important part in "tailoring" the oil but it

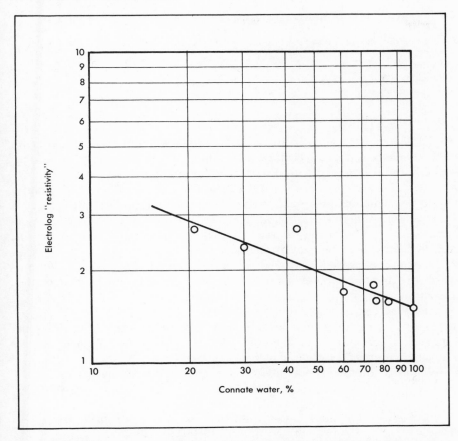

Figure 3–9 Connate water vs electric-log "Resistivity." From Bruce and Welge, Production Practice and Technology, 1947, p. 173, Fig. 10.

is thought that original oil formation is confined to asphaltic and naphthenic compounds with low heat and pressure parameters.

A bacterial decomposition of organic matter is perhaps the most plausible, causal force in creating oil. Most of the bacteria are capable of living in the absence or presence of free oxygen. Table 3–1 developed by Levorsen from Zobell's "Influence of Bacterial Activity on Source Sediments" shows how efficiently bacteria can convert organic remains with increasing loss of oxygen (from aerobic to anaerobic action).

According to Zobell in "Microbial Transformation of Molecular Hydrogen in Marine Sediments with Particular Reference to Petroleum," quite

TABLE 3-1   CHEMICAL NATURE OF CRUDE OIL AND ITS ENVIRONMENT[47]

Type of Material	Carbon, %	Hydrogen, %	Oxygen, %	Nitrogen, %	Phosphorous, %
Marine sapropel	52	6	30	11	0.8
Recent sediments	58	7	24	9	0.6
Ancient sediments	73	9	14	0.3	0.3
Crude oil	85	13	0.5	0.4	0.1

a bit of hydrogen may be liberated by fermentation of organic matter in the absence of free oxygen. It is also thought that bacteria plays a vital role in the solutions of carbonates through organic acids.

Oil has very complex characteristics which are evident from the many products which can be made from it. When a saturation sample is placed in the core-analysis retort and heated to the required 1,200°, a major portion of the oil in the sample is "cracked". This means that the oil in part or as a whole changes its chemical structure—the carbon and hydrogen atoms rearrange themselves in accordance with the change in temperature and pressure. When the carbon-to-carbon bonds are broken, the resulting compounds will have lower boiling points.

Polymerization is the reverse of cracking.

Isomers are substances of the same composition that have different molecular structure and therefore different properties. For example, normal butane and isobutane are both C_4H_{10}.

Table 3-2 shows the similarities and differences of several gases.

Saturated hydrocarbons which include all the parrafins, are those in

TABLE 3-2   PHYSICAL PROPERTIES OF SOME OF THE MORE COMMON HYDROCARBONS OF THE PARAFFIN SERIES.[38]

Name	Chemical Formula	Molecular Weight	Boiling point (°C.) at normal conditions	Critical Temp. °C	Density Gas (air = 1)	Density Liquid (water = 1) sp. gr.
Methane	CH_4	16.04	−161.4	−82.4	0.554	0.415 (−164°)
Ethane	C_2H_6	30.07	− 89.0	32.3	1.038	0.54 (−88°)
Propane	C_3H_8	44.09	− 42.1	96.8	1.522	0.585 (−44.5°)
n-butane	C_4H	58.12	0.55	153.1	2.006	0.601 (0°)
Isobutane	C_4H_{10}	58.12	− 11.72	134.0	2.006	0.557
n-pentane	C_5H_{12}	72.15	36.0	197.2	2.491	0.626
Isopentane	C_5H_{12}	72.15	27.89	187.8	2.491	0.6197
n-hexane	C_6H_{14}	86.17	68.75	234.8	2.975	0.6594
Isohexane	C_6H_{14}	86.17	60.30	228.0	2.975	0.6536

which the valence of all the carbon atoms is satisfied by single covalent bonds. This type of structure is very stable. Unsaturated hydrocarbons are those where the valence of some of the carbon atoms is not satisfied with single covalent bonds so they are connected by two or more bonds which make them less stable and more prone to chemical change.

In exploratory drilling, one is not usually concerned with the various breakdowns of the oil, particularly at the drilling site. What one is most concerned with is the type of oil found and its gravity. Fig. 3–10 shows the relationship of the four types of oil.

The paraffin series begins with methane (CH_4), but its basic formula is C_nH_{2n+2}. Pentane to pentadecane are liquid and the chief constituents of uncracked gasoline. Its higher members become waxy solids. In a given bore hole, the wax may clog the pore space next to the hole as gas expands and cools under inordinant producing pressures.

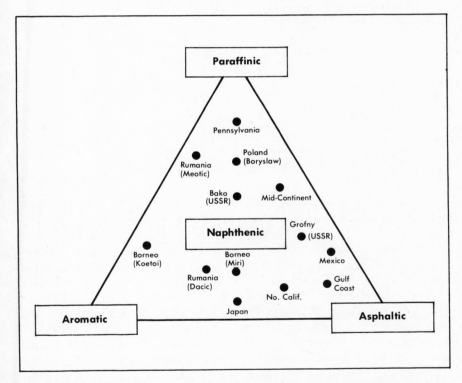

Figure 3–10 From The Chemical Technology 54, W. A. Gruse and D. R. Stevens, Copyright 1942. Used with permission of McGraw-Hill Book Company.

The naphthene series (C_nH_{2n}) is identified by having single covalent bonds, being a closed-ring series, and is saturated. A crude oil with a high naphthene content is referred to as an asphalt-based crude oil whenever there are complex asphalt members with it.

The olefin series is made up of the same parts but is unsaturated and open-chained.

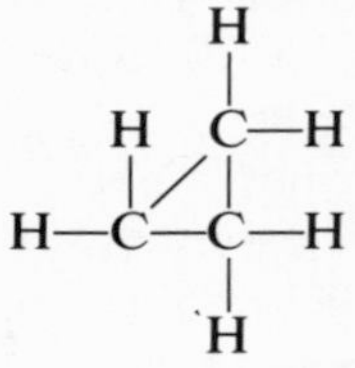

$CH_2CH_2CH_2$
Mol. Wt. 42.08
B.P. −34.4°C

Naphthene (cyclopropane)

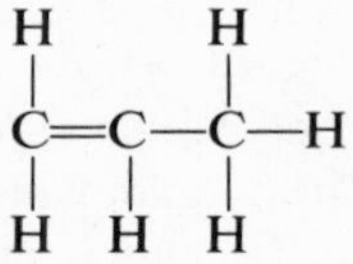

$CH_2 \cdot CH \cdot CH_3$
Mol. Wt. 42.08
B.P. −420°C

Olefin (propylene)

The aromatic series (C_nH_{2n-6}) is an unsaturated closed-ring series with a strong aromatic odor.

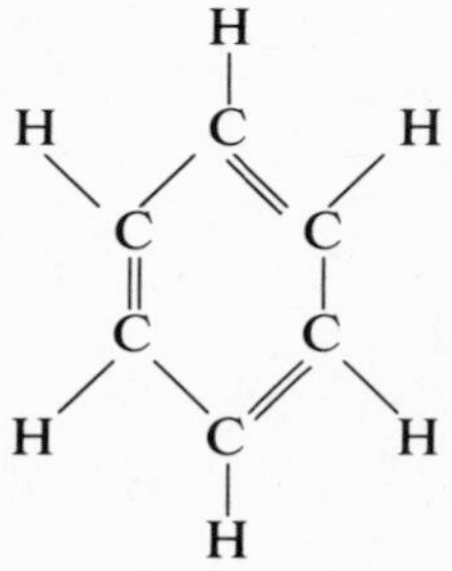

Benzene C_6H_6

Mol. Wt. 78.11
B.P. 80°C.

Asphalt is not a series by itself. Asphalts are highly viscous to semisolid, brown-black hydrocarbons of high molecular weight usually containing a lot of sulfur and nitrogen, which are undesirable components, and oxygen. Asphalts are closely related to the naphthene series and because of their high nitrogen and oxygen content they may be considered juvenile oil, not fully developed.

The aromatic-naphthene based crudes are usually associated with limestone and dolomite reservoirs such as those found in Persia, the Arabian Gulf and Borneo.

Since there are so many types of oil, each with a wide range of specific gravity, an arbitrary nonlinear relationship was developed by the American Petroleum Institute (API) to classify crude oils by weight on a linear-scaled hydrometer. The observed readings are always corrected for temperature to 60°, by using a prepared table of standard values, see Appendix 2

$$\text{Degrees API} = \frac{141.5}{\text{Sp. Gr. at } 60°\text{F.}} - 131.5$$

The general refractive index range for oil is 1.39 to 1.49. The heavier the crude, the higher the refractive index and the lower the API gravity. This can be measured with a refractometer or by the same methods used in optical mineralogy with reference gravity oils.

Fluorescence of oil is measured by its color under ultraviolet light. The cored sample should be placed as quickly as possible under ultraviolet light since fluorescence of oil subsides with evaporation and the activity of "live" oil decreases. If possible the whole core should be passed under UV light to determine the fluorescent color and the pattern of oil in place in the cored interval.

When possible, pictures should be taken of the core showing the fluorescence. These are very useful when accompanying reports to the head office which may be hundreds if not a few thousand miles away.

2° − 10° API	Nonfluorescent to dull brown
10° − 18° API	Yellow-brown to gold
18° − 45° API	Gold to pale yellow
45° − above	Blue-white to white

The flash point of crude oil is measured by heating the oil in a container and passing a lighted match across its surface. When flashes of short duration appear, the temperature of the oil is recorded from a thermometer which has been registering the temperature increase as the oil was heated.

Needless to say this operation should be done under controlled, safe conditions.

The burning point is the lowest temperature at which crude oil will ignite and maintain a flame. The color of these flames should be described since they too are indicative of characteristics interesting to the company production department. The flash point and burning point, among other things indicate the dangers of running drill-stem tests of the oil zone. Usually, not enough free oil is obtained with a core to make these analyses, but an attempt to light the oil in the core at atmospheric temperature should be made.

Cloud points and pour points are again not measurable from a core but with special preparation of a freezing bath, tests may be made on oil samples from drill-stem tests which generally follow coring a prospective

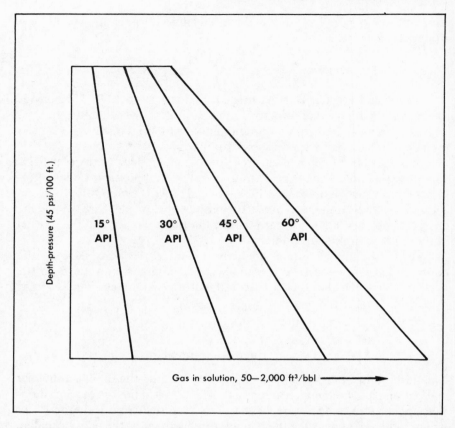

Figure 3–11 The general effect of increasing pressure with depth on the capacity of crude oils to hold gas in solution. From Beal, The Viscosity of Air, Water, Natural Gas, Crude Oil and Its Associated Gases at Oil Field Temperatures and Pressures. Transaction SPE-AIME Vol. 165 Copyright © 1946.

reservoir interval. The oil is placed in a container, preferably a pyrex bottle. A thermometer which can measure temperature variations from −70°F. to 100°F, is inserted in the container to measure the oil temperature as it cools when the container is immersed in the freezing bath. The bath may be dry ice or liquid nitrogen, if available.

The container is removed from the bath occasionally and tilted so the oil will move. The cloud point occurs when the first cloud appears: the temperature should be recorded. The clouds are due to congealing paraffins. Oil without paraffin constituents have no cloud point.

The Pour Point is only a few degrees below the cloud point and it is described as the point when oil will no longer pour from the container. This temperature is recorded.

With a polarizing microscope it is possible to determine with crossed nicols whether oil is Dextrorotary or Levorotary more commonly known as "righthanded" or "lefthanded". The amount of rotation to the right or left is measured on the stage in degrees per millimeter.

The average range of rotation is from 0° to 1.2°/mm.

The importance of these measurements lies in the speculation of the genesis of oil itself, since optically active oils cannot be synthesized by inorganic means. The right or left-hand pattern could be caused by cholesterin-like material and perhaps the direction and degree of rotation will one day indicate the type of matter involved.

Centrifuge analysis is more applicable to drill-stem testing but again is mentioned since DSTs usually follow coring operations. Centrifuge analysis permits the wellsite geologist to determine the percentage of gas, water, and sediment in the oil sample. This analysis must be made quickly if the gas in solution is going to be measured with any near accuracy. Fig. 3–11 shows the relationship of gas in solution at depth for various gravities.

It should be pointed out that most oils increase in API gravity with depth in a given lithologic column with the reason being that younger juvenile oils, heavier with a lower API gravity, have not yet been transformed from the initial formation conditions to higher petroleum members. Two well known exceptions to this pattern are found in the Burgan sands of Kuwait and the shallow sands of the Bibi Eibat field in the USSR where the high-gravity members are found higher up in the stratified column than the low-gravity members.[52]

4

Coring Tools,
Analysis Equipment,
Checklists

It was mentioned in the introduction that the first coring tool appeared in 1908 in Holland. The first one used in the United States appeared some years later (1915) and was a piece of modified drillpipe with a saw-toothed edge for cutting—much like a milling shoe.

Core-analysis equipment has been developed primarily during the last 25 years. The instruments are very functional, accurate, and robust for field work.

The checklist kept by the geologist making an analysis is very important so that time is not wasted and steps forgotten in handling and processing the core. On wells which are far from ready supply stores, the list provided will prove an effective means by which to take stock of the required items to complete a suite of analyses.

4.1 Coring Tools

With the modified drill pipe and saw-toothed cutter mentioned above, the required amount of rock was cored and held by the rotating drilling string. When enough core had been cut, weight was added to the string to press the jagged cutting teeth inwards. This effectively cut the core and held it in place for the trip to the surface. Since there was no proper fluid flow, the cores were usually badly burned and not of much use except as curiosity pieces.

J. E. Elliott in 1921 was the first individual to use the inner barrel and toothed bit successfully. In 1925 the facility had been added to allow exchange of core bits; the inner barrel had been designed to remain stationary; a core catcher had been added.[54]

Since coring was first attempted, many variations of the same principle of operation have been developed. Presently three types of coring tools do over 90% of the coring work. They are the conventional 3-cone roller, wireline (roller and diamond), and full-core diamond cutting tools. The diamond tools are now recognized as the best tools for coring (both ordinary and wireline) and are used almost exclusively today because of the expense of coring, long down-hole durability and reliable cutting and recovery capability.

In Fig. 4–1 (a and b) are shown a diamond-bit core barrel, where the core is retrieved by pulling the barrel to the surface with the drilling string, and a wireline core barrel, where the core is retrieved by using an overshot on a sandline while leaving the bit and drilling string in the hole. The ordinary barrel is made up in 30-ft lengths so 30, 60 or 90 ft or more can be cored at one time. The wireline barrel is 15 ft long.

One special tool adaptation used to obtain cores in soft or fractured formations is the rubber sleeve core barrel as shown in Fig. 4–2.

The side-wall coring device was introduced around 1940 for the purpose of taking samples of prospective formations from holes already drilled to a predetermined depth. The bullets, shown in Fig. 4–3 are lowered in a tool (gun) to the depths required and fired electrically. With the samples caught in the bullets, the gun is reeled to the surface and the samples extracted.

This method of coring physically verifies the formation type shown on the electric log. The size of the plugs is small and therefore there is not enough material to make appropriate porosity, permeability, and saturation checks. The samples from reservoir formations are usually flushed with mud.

Side-wall cores are taken to examine the formation of interest, to see what fluids are present and to check for any paleontological evidence. They should never be taken with the intention of evaluating extensively the porosity, permeability, and saturation characteristics of the rock.

Side-wall coring is an inexpensive method for substantiating a possible prospect's requirement for more detailed testing. It is best used in exploration wells to verify electric-log interpretation of lithology and in confirmation wells to check for any formation changes likely to affect production.

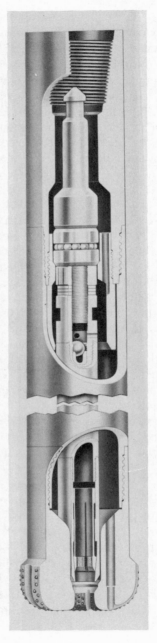

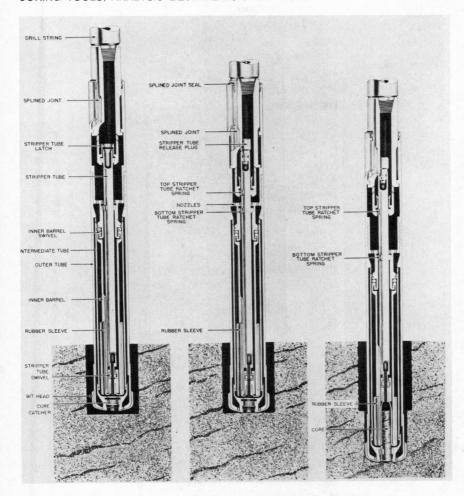

Figure 4–2 Rubber-sleeve core barrel. (1) Lowered to bottom of hole; (2) ready to cut core; (3) after cutting 2 ft of core. This tool is especially useful in soft or fractured formations. (Courtesy Christensen Diamond Products.)

Figure 4–1 Commonly used core barrels. At left (a) is the 250 P Model which can be made up in 30-ft sections. It features a safety joint to allow pulling the inner barrel with the cores in case the outer barrel becomes stuck. At right is a wireline core barrel that permits continuous coring without pulling the pipe. The core-holding tube is held in place by mud pressure. (Courtesy Christensen Diamond Products.)

Figure 4–3 Side-wall coring tool.

4.2 Analysis Equipment

The most widely used equipment for both field and laboratory core analysis is designed by Ruska of Houston Texas. Discussions and explanations conducted in this handbook will be concerned primarily with Ruska equipment.

The basic instruments are the porometer for porosity measurements, the permeameter for permeability measurements, and the saturation retorts for saturation measurements.

The liquid permeameter and mercury-injection capillary pressure apparatus, as shown in Figs. 4–4 and 4–5 are not discussed in detail since this equipment, unfortunately, is not ordinarily found at the wellsite. Where it is used, the individual who has learned to operate the porometer and permeameter will have no difficulty in adapting to the necessary techniques required for operating the additional equipment. Other equipment used to process the core will be discussed.

Porometer. The universal porometer as shown in Fig. 4–6 consists of a 100-cc volumetric mercury pump to which a pycnometer vessel is attached. The pycnometer will take core plugs $1\frac{1}{4}$ in. long and $1\frac{1}{2}$ in. in diameter. The pycnometer has a breech-locked lid which utilizes an "O" ring for pressure sealing. A needle (petcock) valve in the pycnometer lid

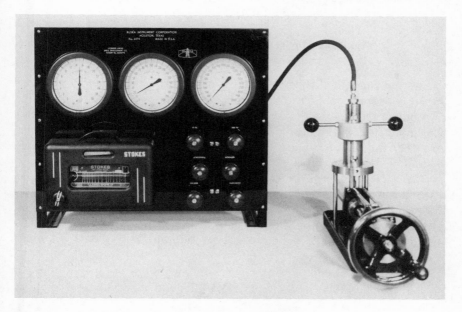

Figure 4–4 Mercury-injection capillary pressure apparatus.

Figure 4–5 The liquid permeameter.

allows opening and closing the sample chamber to the atmosphere when
the lid is breech-locked to the top of the chamber.

Volumetric readings to an accuracy of 0.01 cc can be measured and
read directly with the porometer. The universal porometer may be used
for either the mercury-injection porosity method or the Boyle's Law air
method.

The 0.01-cc accuracy already mentioned can be improved on with the
Boyle's Law-type air-injection method which is described in Chapter 7.
Correction factors are applied in both the mercury-injection and air-
injection methods to limit the errors inherent in the porometer, and the
changes in temperature and atmospheric conditions.

Metal Reference Plugs The metal reference plugs are used in the air
porosity measurement technique. Three plugs of different sizes such as
3 cm³, 4.5 cm³, and 5.5 cm³ are used as reference volumes to plot a ref-

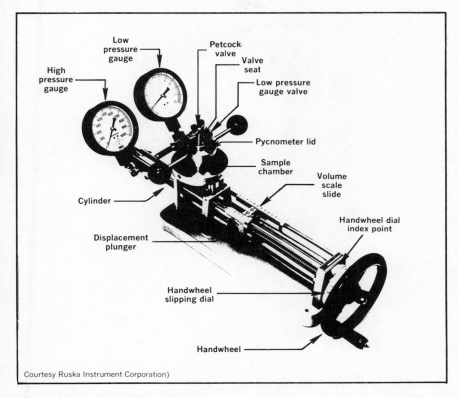

Courtesy Ruska Instrument Corporation)

Figure 4-6 Universal porometer.

erence line on a graph as shown in Appendix 2. This method described in Chapter 6 is a more accurate method than the Boyle's Law method in that the different-sized reference-volume plugs ensure the accuracy of the porometer over a range.

It is a matter of having four reference points with the air-injection method and only one reference point with the Boyle's Law method. Accuracy is increased to 0.001 cc.

Permeameter. The permeameter as shown in Fig. 4-7 utilizes a modified Fancher-type rubber holder called a "sleeve" for drilled or sawed samples and a metal bushing adapter for shaped samples which are mounted in wax. Where cubes or square-ended samples are to be examined, a suitable sleeve is available.

The permeameter is designed to measure the flow of air through a core

Figure 4–7 Gas permeameter.

sample by forcing air into the sample. The amount of air pressure applied to the sample is measured by a floating ball in any one of three different-sized flow-meters and a pressure control valve. The registered air pressure seen on a particular flowmeter being used, in conjunction with a specific air pressure at a known air temperature that is throttled against the sample (controlled by the air pressure dial), yields a viscosity factor which is applied to a prepared graph as shown in Fig. 4–8. This graph is made for each individual machine and the serial number on it should be the same as the one on the machine.

Where an air compressor is used as the air supply, its pulsations should not be allowed during analysis; make measurements only with a fully charged air tank. The air must be dried and the pressure reduced to 20–40 psi before entering the permeameter. The range of the permeameter is from less than 0.5 to 3,000 md.

Saturation Retort. The retort apparatus consists usually of an insulated oven, a water-filled condenser jacket and a receiver. Ruska make four different types of stills utilizing water and electrics, water and gas, air and

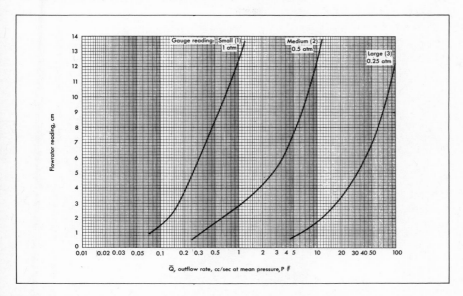

Figure 4-8 Calibration curves for gas permeameter flow meters.

electrics, and air and gas, but the one first mentioned is the type most commonly found on drilling sites.

The instrument is normally mounted in banks of two, six or eight, depending on the amount of work to be done and the speed with which it must be accomplished. The retorts, referred to as "stills", extract the fluids from the rock samples which have been placed in the retort sample cylinders. Approximately 125 g of sample is an ideal amount for the retort. In a suite of analyses, accuracy is increased in the resulting values of saturation if the samples retorted are approximately the same size. See Fig. 4-9.

Since the retorts heat up to a temperature of 1,200°F., it is impossible to take the retorted water or oil to ascertain salinity and gravity values. Distilled water and fractioned oil are received from the retorting process to which table values are applied to correct the amounts to "approximate actual" amounts contained in the rock during original surface conditions. There is much debate on the validity of the values obtained with current retort methods, but suffice it to say that the results will coincide reasonably well with porosity measurements and electrical log data.

The water-cooled stills have closed condensers with flow controls for the water so that the operator can adjust the cooling rate in order to assure fairly complete recovery of the distillate. The light hydrocarbons and

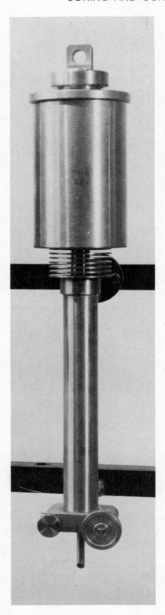

Figure 4–9   Fluid-content still.

water which are evaporated first are condensed at a low temperature while the heavier, more viscous distillates, which may cling to the walls of the condenser tube at low temperature can flow through the condenser at progressively higher temperatures.

The temperature is controlled by first circulating water through the condenser jacket, then stopping the circulation and finally draining the condenser jacket of water. The times for the three phases of operation are:

1. Up to 15 min with water circulation.
2. From minute 15 to minute 25, stop water circulating.
3. From minute 25 to minute 40, retort with drained condenser.

Diamond Core Drill. The core-plug drill is a hand drill mounted on a bracket to facilitate the cutting of plug samples. Normally a $\frac{3}{4}$-in. diamond core bit is used.

It is best to use air as the cooling medium around the bit but water may be used if there is no problem of changing the mineral constituency.

When air is used, the cutting should not be carried out in a confined space as dust is created. An extraction fan with hood solves the dust problem.

When the core is friable and will not remain consolidated during cutting, compressed gas such as CO_2 can be used to freeze the wall of the plug as it is cut. Normally the plugs are cut to a length of $\frac{1}{2}$ to 1 in. with the average size being $\frac{3}{4}$ in.

This cutting procedure cannot be used if the plug is unconsolidated, in which case it must be cut by hand using a knife.

Soxhlet Extractor. This instrument is used to extract all the fluids from the samples used for porosity and permeability measurements.

It is best to have two extractors for extracting fluids to speed up the analyses. Basically, the operation consists of vaporizing a solvent such as toluene, filled halfway in the bottom flask, which permits the greatest area to be exposed for surface vaporization. The vapor passes to the middle chamber after being condensed at the top chamber by a circulating cold-water jacket. The condensed solvent gradually fills the middle chamber and eventually is syphoned back into the lower chamber.

In the repetition of the cycle, the water and oil from the sample being cleaned remain in the bottom chamber since toluene vaporizes more readily than water or oil.

Core Boxes. The boxes used to carry the cores should accomodate 3 ft or more of core. This is the easiest length to handle especially when extracting the core from the barrel. It is rather easy to get 3-ft bits of core out of the barrel without breakage.

The system for marking the boxes varies from company to company and operation to operation, but generally the boxes should be marked on the end starting with the core number and the box number written underneath starting with Number 1. Number 1 is the bottom of the core and therefore the first out of the barrel.

On the side of the box an arrow is used to show the orientation of the core in the box and sometimes a "T" for top on one end and a "B" for bottom on the other end.

After the core has been received and examined, the top and bottom depth of the core in the box may be marked on each end of the lid before it is nailed to the box.

API Oil Gravity Hydrometers. There are two hydrometers necessary for making specific gravity measurements. One measures from 10 to 45° API and the other from 45 to 90° API.

Weighing Scales. These, which measure the weight of a sample to 0.01-g accuracy are necessary to insure optimum results in the analysis.

Scales with finer measurement capabilities cannot be used on offshore rigs because the rig movement upsets the balance too greatly.

It is desirable but not essential that the scales can weigh as much as 600 g, in which case the retort sample can be weighed in the retort cylinder. This shortens the procedure and provides more accuracy through less handling of the sample.

Sieves. Sieves are used to sort the sample particles from undesirable ground-up rock particles. Unless the core is a hard carbonate or evaporate, it is possible to make rock particles which will fall through a $1/4$-in.-mesh screen. If this is too small, a $1/2$-in.-mesh screen should be used.

Uniformly sized particles mean a more consistent set of figures through the analysis.

Core Dip. Core pieces for laboratory examination have been put in a variety of containers for shipment. Only two types of containers are commonly used now if the operator wishes to get the core to a sophisticated laboratory in rig-floor condition.

Both core dips seal the core with an impervious coating. The first is a "hot dip" called Core Gel. Its main ingredients include ethyl cellulose or cellulose acetate butyerate, both of which are strippable, pliable plastics. The Core Gel is heated in a pan in which the cored pieces for preservation are dipped. A wire wrapped about the cored piece is used to hang it on the core hanger for drying.

Since hot dip does not work satisfactorily in hot and humid climates another type has been developed by Coroda Chemical, Snaith Goole, Yorks, England, which requires no heating of the liquid. This cold dip has two main advantages in that it dries faster and less plastic fluid is required to make an impermeable coating about the sample. It is not advisable to use twine or string to hold the samples for dipping and drying since it is porous and there is the possibility of leakage from the sample to the atmosphere.

Vacuum Oven. This is a closed, vacuum-sealed metal container which is heated. The heated oven with its connected vacuum line will draw any remaining moisture from the cleansed porosity and permeability samples placed inside.

The alternative drying device is an ordinary cuttings sample drying oven which is not as efficient and allows the moisture of the atmosphere to interfere with the process.

Dessicator. This is a metal container divided into top-and-bottom sections. The bottom section contains chemical particles which absorb moisture.

In the top part is placed the sample(s) to continue drying and cooling after being in the vacuum oven. The top of the container is sealed to the lid with petroleum jelly or thick oil to maintain an air-tight condition.

Marking Tags and Trays. Careful marking helps maintain organization through the many steps of the analysis. Small metal tags stamped with numbers from 1 to 100 should be enough. Each tag is tied to a sample with a thin pliable wire.

The trays are small boxes into which the samples can be placed while waiting to process them through the analysis.

Ultraviolet Light. The "UV box" should be a portable type with a long extension cord. The core is examined under UV light as soon as possible for mineral and oil fluorescence.

The mud on the core surface should be wiped away, not washed away.

Binocular Microscope. The microscope is as useful as the ability of the geologist to use it. One which can magnify through four or five lenses up to 50 power is ideal.

A rotating stage is helpful in observing the character of the pore spaces in the sample.

In most field analyses there is too little emphasis placed on microscope evaluation of cores. The logging geologist should look at a freshly broken piece of core, make a written description, clean the sample with solvent in the extractor and re-examine the sample for any changes.

TABLE 4–1   WELLSITE EQUIPMENT CHECKLIST

Receiving the core; preparation for shipment

Core boxes
Nails
Core hammer
Marking paint & brush
Pocket knife or putty knife to scrape
   mud from core.

Portable ultraviolet light to examine
   core for oil, condensate, or mineral
   fluorescence.
Core dip, hot or cold variety
Wire for hanging core pieces to dry.
Aluminum foil (if core dip isn't used)
Suitable apparatus to hang drying cores.

Preparing Core samples

Diamond core drill
Soxhlet extractor w/stand
Toluene
Rubber tubing
Hotplate
Rubber tubing clamps
Vacuum oven
Dessicator

Marking tags and trays
Forceps
Sealing wax
Salt
Oil or petroleum jelly
Weighing scales
Seive ($\frac{1}{4}$-in. mesh) and sample catchbasin
Mortar & pestal (or rock crusher)

Measurements of core samples

Core analysis master worksheet
Caliper w/vernier scale
Thermometer
Detergent
Permeameter
Porometer
Pycnometer cleaning brush
Saturation retort(s) w/receiving tube
   stoppers
Saturation data sheet

Tempelstick
Retort brush cleaner
Stopwatch
Retort receiving tubes
   (graduated in cc)
Hydrometers
Alcohol
Detergent
Specific gravity (grain density) cylinder.
   (Optional)

Calculations and report preparation

Pad of paper
Slide rule
Appropriate graphs and charts

Pencil and ruler
Pad of core-analysis report forms.

4.3 Equipment Checklist

Good core analysis at the wellsite is impossible without the proper equipment.

Table 4–1 provides a convenient checklist for the wellsite geologist. This list should be checked before each trip to the field.

5

Coring, Receiving, Sampling, and Preservation

It would be impractical — and boring — to explain in strict detail all the approaches to the coring and analysis process. To limit the scope of this work, the discussion will be restricted to the diamond bit, today's most popular technique.

5.1 Coring

Diamond coring methods retrieve longer cores than possible with conventional core bits because they can stay on bottom longer. Core barrels up to 60 ft and more in length are used.

Every method of coring stresses the necessity for a clean hole but it is especially true in diamond coring operations because of the expense of the bit. Each diamond on the bit is continually on bottom, continually doing work. The bit must be kept clean and cool for efficiency, prevention of balling up, and to keep new formation continually exposed to cutting action to prevent loss of diamonds.

Any junk lost in the hole — like bearings and cones — must be fished out rather than forced to the wall of the hole. A junk basket should be run during two or three roller-bit runs prior to coring. A wiper should be around the drill-pipe at all times to guard against tools being accidentally dropped in the hole (an old wiper is best for running in the hole).

Most operators require a substantial circulation period and the running of a magnetic tool to bottom before coring.

The drill pipe is held in tension, as normal, by carrying sufficient drill collars in the string. All core barrels are equipped with a stabilizer positioned directly above the bit. The stabilizers are machined to be from zero to 0.03 in. smaller than core-bit diameter. When they wear so that they are $\frac{1}{16}$ in. under-sized they should be rebuilt or replaced.

Most diamond coring is now done in the full diameter of the borehole. Whether coring is done this way or in a rat hole depends on the full gauge-keeping ability of the rock bits used prior to initiating the coring operation.

It is poor practice to run the same-sized diamond bit as the roller bit with the intention of reaming to bottom successfully. Diamond bits are not designed for reaming for they are relatively weak in shear. Their strength lies in their use under compression between the rock and the bit face.

With reaming, there is a good chance of key-seating the bit since its rotational path is different from that of a three-cone bit. The amount of gauge loss on the roller bits prior to coring is a good measure of the decrease in hole size one must consider when running the coring string.

The way a diamond bit is used governs its effectiveness as a drilling tool The bit cannot be treated as roughly as a lot of roller bits are treated. Burning of the diamonds occurs when the bit face becomes clogged with gummy formation material or the bit is "crammed" without increasing the hydraulic power and fluid circulation.

Diamonds will break off the face of the bit when too much shearing weight is added as evidenced by severe vibration and bounce in the drilling string.

Each diamond core barrel type and size will permit a given maximum fluid flow between the inner and outer barrels. The data for various barrels are given in Table 5–1.

Figure 5–1 is a general graphical representation of drilling weight versus bit diameter. See Appendix A for an explanation and diagrams of the methods used by Christensen Diamond Products to operate their bits.

The assembly of the core barrel is very important to ensure reliable performance. The joint threads should be "hand-tonged" going into and out of the hole. The spinning chain and rotary table should never be used to make up or break apart the threaded joints. The make-up torque should be exactly as the barrel manufacturer recommends. To prevent bending of the barrels, the joining of the sections should be as close to the rotary table as possible. The inner barrel should be made up with chain tongs— the outer barrel can be made up with derrick tongs.

TABLE 5–1 TYPES OF CORE BARRELS (Courtesy of International Association of Drilling Contractors)

Core barrel size & type	Outer tube OD, in.	Outer tube X-sectional area, sq in.	Core dia., in.	Annulus area, sq in.	Fluid capacity, cpm
AMERICAN COLDSET CORPORATION					
Standard core barrels					
$3^5/_8$" × 2" × 40'	$3^5/_8$	3.83	2 or $1^7/_8$	1.08	85
$4^1/_8$" × $2^1/_4$" × 60'	$4^1/_8$	5.07	$2^1/_4$	1.80	140
$4^1/_2$" × $2^1/_2$" × 60'	$4^1/_2$	5.58	$2^1/_2$ or $2^3/_8$	2.65	205
5" × 3" × 60'	5	7.07	3 or $2^7/_8$"	2.25	175
$5^3/_4$" × $2^1/_2$" × 60'	$5^3/_4$	8.25	$3^1/_2$ or $3^3/_8$	3.54	275
$6^7/_8$" × $4^3/_8$" × 60'	$6^7/_8$	10.01	$4^3/_8$ or $4^1/_4$	4.42	345
Maritime core barrels					
$6^1/_4$" × 4" × 60'	$6^1/_4$	16.49	3	3.14	245
$4^1/_2$" × $2^1/_8$" × 60'	$4^1/_2$	7.61	$2^1/_8$	1.80	140
CHRISTENSEN DIAMOND PRODUCTS COMPANY					
$3^1/_2$ × $1^3/_4$ Slim Hole	$3^1/_2$	3.7	$1^3/_4$	1.5	118
$4^1/_8$ × $2^1/_8$ 250 P Series	$4^1/_8$	5.1	$2^1/_8$	1.8	141
$4^1/_2$ × $2^1/_8$ Marine Series	$4^1/_2$	7.6	$2^1/_8$	1.8	141
$4^3/_4$ × $2^5/_8$ 250 P Series	$4^3/_4$	6.7	$2^5/_8$	2.1	164
$5^3/_4$ × $3^1/_2$ 250 P Series	$5^3/_4$	9.2	$3^1/_2$	2.6	204
$6^1/_4$ × 4 250 P Series	$6^1/_4$	10.1	4	2.9	227
$6^1/_4$ × 3 Marine Series	$6^1/_4$	16.5	3	3.1	245
$6^3/_4$ × 4 250 P Series	$6^3/_4$	13.1	4	5.0	387
8 × $5^1/_4$ 250 P Series	8	15.8	$5^1/_4$	3.8	295
RUCKER-HYCALOG					
$4^1/_2$" × $2^3/_8$" × 60'	$4^1/_2$	5.22	$2^3/_8$	2.38	235
$5^3/_4$" × $3^1/_2$" × 60'	$5^3/_4$	8.25	$3^1/_2$	3.50	280
$6^7/_8$" × $4^3/_8$" × 60'	$6^7/_8$	11.16	$4^3/_8$	4.32	340
$6^7/_8$" × $3^1/_2$" × 60'	$6^7/_8$	19.40	$3^1/_2$	3.54	280
WILLIAMS DIAMOND BITS					
7" HW	7	20.76	$3^1/_2$	3.53	275
$6^1/_2$" HW	$6^1/_2$	15.46	$3^1/_2$	3.53	275
$4^3/_4$" HW	$4^3/_4$	10.65	$1^7/_8$	1.878	146
$4^1/_8$" HW	$4^1/_8$	6.29	$1^7/_8$	1.878	146

The fluid capacity in the above tube is based on maximum fluid velocity of 25 fps in the annulus between the outer and inner tubes.

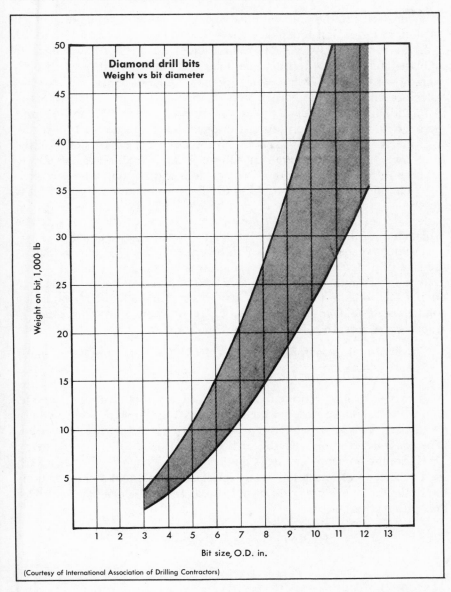

Figure 5–1 Recommended operational range for diamond drill bits. Courtesy of International Assoc. of Drilling Contractors.

When going in the hole with the coring assembly, hold up about 2 ft. off bottom (after first touching bottom) and adjust the pump rate to deliver the required mud volume for the diamond bit, not exceeding the amount that can pass between the inner and outer barrels. A good optimum flow rate is between 5.3 and 7.0 gpm/sq. in. of hole area.[57] No two holes are alike and the correct combination of weight on bit, rotation, and hydraulics can be found by the experienced driller. The largest single factor in improving the drill rate with diamond bits is the rotation speed.

After the pump rate is fixed to suit the bit and the consistency of the formation, the bit is lowered to bottom without rotation and 5,000 to 8,000 lb. is applied to ensure it is on bottom and not on cuttings or cavings. The bit is picked up, rotation of 40–50 rpm started, and the bit is slowly lowered back to bottom and weight applied gradually as the bit pattern is bedded in. When the bit pattern has been established to the satisfaction of the driller, normal drilling weight can be applied.

Drilling weights are increased in increments of 2,000 lb. As long as the rate of penetration increases, additional weight can be added.

The pressure drop across the face of a diamond bit has a tendency to lift the face off the bottom. This "hydraulic pumpoff force" should be compensated for through the addition of bit weight. A nomogram in Appendix A corresponds to the following formula.

$$\text{Hydraulic pumpoff force (lb)} = 0.785 \times \text{bit pressure (psi)}$$

$$\times \,(1.2 \times \text{bit diameter (in)} - 1.2)$$

With the bit drilling (coring), the pump strokes should be the same as before coring was started, but the pump pressure should now be 200–275 psi greater. This represents the pressure drop across the face of the bit. (The pressure drop may be 100 to 1000 psi).

Hydraulic horsepower across the face of the bit can be varied according to the type of formation being drilled. In shales and soft formations the optimum is 2–3 hp/sq in. of hole area. In harder drilling, 1.5 to 2.5 hp/sq in. is desirable.[58]

$$\text{Hydraulic horsepower} = \frac{V \times P}{1,714}$$
$$\text{(hhp)}$$

where:

V = volume in gpm
P = pressure loss, psi

$$\text{hhp/sq in.} = \frac{\text{hhp}}{A}$$

where:

hhp = Hydraulic horsepower (above)
A   = Area of the bit (sq in.)

Pressure drop should be watched closely. Any deviation from it during a coring operation means that something abnormal has happened and core recovery is in jeopardy. If the pump pressure and torque fluctuate for no understandable reason, the bit should be raised off bottom and restarted. Core blockage or formation fracturing are the probable causes. If the problem persists, attempts to drill at an increased weight for a couple of inches should be made. If neither method solves the problem, the bit must be pulled.

If only the pump pressure increases inordinately, lift the bit off bottom. If, by taking the bit off bottom, the pressure is relieved, an "O" ring groove in the bit face has formed which causes restricted mud flow and will result in a severe loss of diamonds rather quickly.

If there is a decrease in the penetration rate accompanied by a decrease in pump pressure and/or torque, it means the formation is being drilled rather than cored. The causes of this are several but most often it is caused by a blockage in the core barrel by a piece of core which is fractured and becomes stuck.

Drilling weights with diamond core bits are somewhat less than those with roller bits. A good optimum weight ranges from 750 to 1,000 lb/sq in. of hole area. The weight must be applied uniformly. See Appendix A.

Rotary speed is varied to find the optimum speed to suit the mud flow and the drilling weight. Most drillers talk to the mud logger before coring to learn about the formations plasticity, hardness, brittleness, and fracturability. Often this informal chat contributes to a successful operation which otherwise might just be another "cramming" session trying to get the fastest drill rate while coring. Rotary speed of 100 rpm is average but diamond bits can operate up to 1,000 rpm when down-hole turbines are used.

The fastest drill rate is not the optimum drill rate. As already mentioned, pump pressure plays an important part in coring, especially when coring friable sands and fine calcarenitic limestones which can be washed away by hydraulic action at the bit face. The pumping rate (pump pressure) must be selected with respect to the formation as well as the drilling bit.

When a connection is made or the bit must be taken off bottom for any reason, stop the rotation and shut off the pump; raise the core barrel until the weight indicator shows the spring has gripped the core and broken it. No more than 20,000 lb should be required to do this.

If the core will not break, circulate with drilling pump pressure and hold the strain on the core. When the core has broken, raise the bit 10 ft and slowly lower back to where bottom should be, as marked on the kelly to make certain the core has been picked up.

After making the connection, put the bit back on bottom without rotating and add about 50% more weight than used when coring . . . this will allow the new core to enter the barrel. Raise the drill string until the recommended drilling weight is on the weight indicator and start the pump and rotary to continue coring.

Where it is known that the formations are fractured, it is good practice when making a connection to lock the rotary table so the fractures will mesh and not crumble the core. This can cause jamming and an unnecessary trip.

The mud-logging geologist will be primarily concerned with the drill rate and gas recorded while the core is being cut but perhaps more important, he must continue to collect cuttings samples at the shaker for the case where there is incomplete or negligible core recovery.

The drilling rate during coring will vary with the type of bit, the pump pressure, weight, rotary rpm, and formation drilled. The gas curves recorded by the mud-logging instruments will be severely damped since there are fewer cuttings and drilling is slow. Basically, until the core is brought to the surface, the logging geologist cannot help but feel that he has been "logging blind" since what little sample that does show up as cuttings on the shaker may come from the wall of the hole, scraped away by the stabilizers.

5.2 Sample Receiving

As the barrel with the core is brought out of the hole, arrange the core boxes on the drilling rig floor, out of the way, but in the order in which they will be filled when the core is received. The drilling crew has several steps to complete before they are ready to help with removal of the core from the inner barrel. Explain to the driller how far off the floor to lift the barrel (2 or 3 ft) so as to get the maximum whole section which will fit into the core box. Be certain, on receiving the core that you put the proper end of the core in the box as it is marked.

Do not put your hands under the core as it is slid out of the barrel . . . a natural reaction to stop broken pieces from shattering and being lost . . . as it may cause severe injury. The inner barrel should never, unless absolutely necessary, be put out the V door with a core inside. It will bend or break at the threaded joint.

Guide the core out with the palms of gloved hands and when the proper amount is extruded from the barrel, hit the core with the core hammer to break it off. Do not let it fall to the rotary table since it will break into two or three more pieces.

The core is best removed gently from the barrel but sometimes solid blows with a light sledge hammer on the barrel are necessary to jar the core to the bottom.

Do not assume that roughnecks automatically know how to receive a core from the barrel. Only the experienced ones do. Whether roughneck, geologist or somebody else, the person working the core grips should have received cores previously.

Frequently pieces in the barrel will slide suddenly and unexpectedly down the barrel with some force. Careful attention by the person using the core grips can avoid a lot of core breakage. The core is not completely out of the barrel until a metal slug, called "the rabbit", appears.

If a part of the core remains lodged in the barrel it must, as a last resort, be pumped out after first laying the barrel on the floor or on the catwalk.

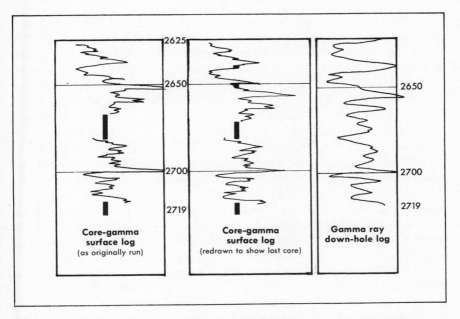

Figure 5-2 Comparison of gamma ray logs makes it possible to locate points where core material was lost. From The Fundamentals of Core Analysis, Core Laboratories, Inc., p. 114.

Measure accurately the amount of core recovered. If the recovery does not equal the drilled interval, it must be assumed that the missing portion was lost at the bottom unless there are circumstances indicating otherwise. The lost portion of the core and any circumstances surrounding the loss should be noted in the final report.

A method not widely used has been developed by Core Laboratories to examine the gamma radiation of a core which, when compared with the gamma ray logs run in the open hole, makes it possible to pinpoint the exact cored zones which have been lost for one reason or another. See Fig. 5-2.

Scrape the mud or, if still wet, wipe it off the face of the cores when in the boxes. Make an inspection of the core suitably detailed to satisfy the objectives of the operator. Select pieces at appropriate intervals for core analysis. The remainder of the core may be washed to simplify the descriptive part of the core report, but only if there will be no further sampling required at the rig or in a laboratory.

5.3 Sampling Guidelines

If the oil company has not specified at what interval to take the samples from the core, or if they have left it to the logging geologist's discretion, samples should be taken which best describe the whole cored section of interest.

For example, do not cut a plug through a single vug to show porosity unless vugs are a firm characteristic of the section. Another case might be where the core was started when drilled cuttings were limestone and when the core is recovered it is learned that the formation had again become anhydrite. Rather than do an analysis on anhydrite, it is advisable to call the operating company for further instructions. It is wasted time and money to analyze the evaporites in the field.

In fairly thick formations of sand, sampling may be done every 2 ft for an accurate analysis. In carbonate formations, dipping formations, and cross-bedded formations, it is advisable to sample every foot or less since the constituency of the formation can change radically throughout the section.

In formation horizons where there are alternately thin sand and shale members, it is necessary to sample every sand member which is over 2 in. thick to obtain reasonable results.

A thin reservoir member with variable lithologic properties should be sampled to ascertain the properties of the top, middle, and bottom of the particular member.

Saturation Sample. Preparation of each sample for saturation measurements must be done quickly since the fluids are continuously evaporating from the sample. The sample should be taken from the center portion of the core where the drilling fluids have affected the saturation the least. Break it up with a mortar and pestal or in the case of harder rocks, pound the sample until there are enough particles which will pass through a $1/4$-in. seive.

Catch the particles in a basin which is attached to the screen. It is wise, when available, to attach a $1/8$ or $1/16$-in. screen between the $1/4$-in. seive and the basin. This allows the dust and small grains to pass through to the basin while retaining the $1/4$-in. particles in the narrower meshed screen. This will avoid the possible clogging of the retort when the sample is heated and the fluids flow to the receiving tubes.

Sometimes the rock is too hard to crush to the desired size so particles larger than $1/4$ in. must be acceptable. The size of the retort sample cylinder will dictate how large the particles may be.

It must be remembered that uniformity of particle size throughout the entire analysis is important. Ideally, one should have a rock crusher which will make uniformly sized particles rapidly with a minimum amount of disturbance to the contained fluids.

When weighed, the sample should weigh between 75 and 150 g. Place the sample into a sample receiving cylinder, mark it, and put it in an appropriate place, ready for processing, if several samples are going to be measured at one time.

Porosity Sample. With the mercury-injection porosity-measurement method it is necessary to cut a plug which will fit conveniently into the pycnometer of the porometer. Again it is necessary to use a sample taken from the inner part of the core away from the wall.

The plug may be taken from one of the pieces initially broken off for the saturation sample. It should be shaped to be approximately $3/4$ by $3/4$ in. which is accomplished with a knife in soft formations or chipping with a hammer in more consolidated formations.

It is important that the sample is freshly cut and not carved with a grinding action since this will close the available pore space.

Very unconsolidated sands which are cemented together only by the oil in place will fall completely apart when the oil is dissolved in the extraction process. Only sophisticated techniques using special sample-holding devices with fluid and air-permeable mediums will allow measurements of such rocks and even then, it is difficult to guarantee that the grain orientation will remain constant. In such sands the grains are usually well rounded and porosities of 35–40% can be conservatively assumed.

With the air-injection porosity-measurement method the sample which is cut for permeability is the same sample used for porosity measurement.

Immediately after the sample is shaped or the plug is cut, weigh the sample and prepare it for fluid extraction in the extractor. Fasten a metal tag to the sample.

Permeability Sample. Where a core is not fractured, whole cores, full diameter cores, and plugs from cores will yield almost exactly the same results. Very often it takes a bit of judgement in deciding just how to cut the permeability plug sample, so a discussion on this point is well worthwhile.

Normally in grossly bedded formations there is no problem and the side of the core is penetrated by the diamond drill. The use of air or water as a lubricating agent is common practice, but water is not recommended since it can alter the pore structure. The use of CO_2 as a lubricating agent is becoming more common now since it prevents overheating of the plug and eliminates closing of pore spaces and loss of fluids in the action.

When the plug is cut to a convenient length, about $3/4$ in. long, make certain that both ends of the plug are freshly broken and do not contain the face of the outer core wall. It is helpful if the ends are relatively perpendicular to the sides of the plug in the interest of making accurate length measurements. This measurement is very important for an accurate analysis. If not perpendicular it is necessary to take several measurements with the caliper and use the average length for calculations.

If the plug is to be used also for porosity measurement as with the air-injection method, it must be weighed immediately. Fasten a metal tag to the plug.

If the core is unconsolidated and cemented together only by the oil in place, it breaks up into unusable fragments. When this occurs it is necessary to cut a sample from the core as done with the mercury-injection method porosity sample. After cutting, shaping, measuring, weighing when necessary, extraction of fluids, and drying, the sample must be set in sealing wax.

Place the sample on an adequate mound of salt and place the inner metal-sleeve core holder around it. Melt the sealing wax to fill the space between the sleeve and the sample. Take the sample from the salt and remeasure the length and diameter of the sample—i.e. the thickness of the sealing wax and the diameter of the sample exposed on each end.

There are no hard and fast rules about sample cutting but the more complicated the bedding of the rock, the more complicated must become the analysis. If the core has dipping beds, the plug is cut along the dip which assumes that the permeability along the strike is identical with that along the dip of the beds. This is not necessarily so since capillary pressure and/or the hydrodynamic pressure created by migrating fluids can cause a linear permeability in a given direction of movement. To check if the permeability along the strike is the same as that along the dip, it is necessary to make two horizontal permeability measurements — one taken along the strike, the other along the dip. This is recommended but it is up to the operating company to decide whether or not it requires this information.

Vertical permeability is measured perpendicular to the strike of the beds and parallel to the side of the core. The core-hole deviation from the vertical should be known and recorded on the analysis sheet.

With the proper cutting tools, it is possible to make cubic plugs which can be oriented and facilitate the measurements of two horizontal and one vertical permeability from a single sample.

5.4 Core Preservation

Before taking the plugs and samples needed to make the core analysis, it may be required to preserve parts of the core for further laboratory study by the restored state method. How much, and which parts of the core most be preserved will depend on oil-company policy and instructions. The material used to preserve the core will depend on what experience indicates is the most satisfactory for the rock in question. Some materials used are:

Air-tight cans. No absorbent packing material should be in the cans. The core piece should very nearly fill the inside of the can. Wrap core piece in foil for travelling before inserting in can.

Sealed aluminum, steel or plastic tubes. Drain pipes with suitable "O" ring-sealed end connecters slightly larger than the core diameter are ideal. Again the core piece must be wrapped in foil for transport to avoid shattering or wear.

Plastic bags. The minimum of air is allowed in the bag. Proper sealing is not easy and there is a chance of a sharp edge breaking the plastic. Where plastic is used, the wrapped pieces should be placed in a container.

Freezing. This method uses dry ice which consolidates the core very quickly. It is useful where the laboratory receiving the cores is nearby. Care must be taken in thawing the cores to avoid atmospheric condensation on the core. Thawing which is done slowly will cause fluid redistribution in the core.

Plastic Coatings. Several plastic wrappings which leave a coating on the core of about $1/16$ in. have been developed. The type coating used is the one that best suits the environment in which the dipping is done. It is the best material to preserve a core which must be shipped over long distances. It is normally made of an ethyl cellulose mixture or cellulose acetate butyerate plastic. The latter works extremely well in hot desert climates such as that found in Libya.

Considering the plastic dips (one is hot, the other is cold), wipe off the mud cake surrounding the sample. If the dip sets immediately, one end should be hand-held while the other is dipped; the sample is reversed and re-dipped to completely encase it. Where a drying period is required to harden the plastic, use a piece of soft wire wrapped about the core and twisted to hold it for the dipping operation. Never use string or twine as it is porous. After dipping hang the core on a core hanger using some bent welding rods as hooks.

Ideally, the cone-preservation step should be accomplished at the same time as the saturation samples are selected.

6

Measurements
and
Calculations

This chapter deals with the measurements of the samples and the calculations of the collected data.

Many of the measurements of porosity, permeability and saturation will be conducted simultaneously in practice, but the subjects will be dealt with separately in the discussion to avoid confusion. The discussion will closely follow the worksheet in Fig. B–3, Appendix B.

6.1 Retort Saturation Procedure

Ruska Saturation Stills are most commonly used for determining oil and water content of cored samples. By subtraction from the whole, the amount of gas contained in the sample can be determined. The sample is heated in a retort which drives off the water and oil. The vaporized fluids are condensed and collected in a graduated cylinder.

(a) Measurements. This is perhaps the most time-consuming of all the measurements to be made. A bank of six or eight retorts is the limit with which one operator can efficiently work in the field. There should be on hand twice as many sample cylinders as there are retort stills to be used.

The retort stills are heated for 30 to 45 min. or until a mark can be made on the surface of the heating chamber with a tempil stick. The internal temperature is then approximately 1,200°F.

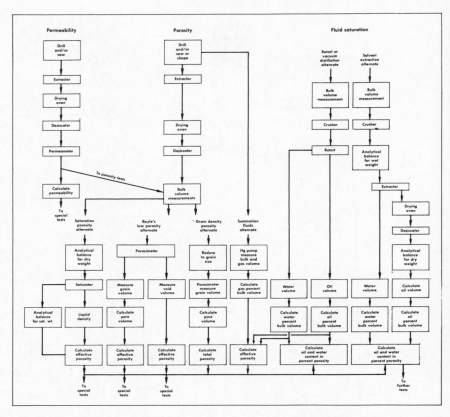

Figure 6-1 Flow sheet of routine core analysis from RP40 "Recommended Practice for Core Analysis Procedure," American Petroleum Institute.

Check that the receiving tubes to be used are dry and put one or two drops of an emulsifying agent in the tubes to insure separation of oil and water during the retorting.

Place the receiving tubes in position to receive the fluids. Regulate a stream of water to flow through the condenser by turning the right-hand knob on each still counter-clockwise. Make certain that the push-pull condenser valve on the left side of each still is pushed to the closed position.

(1) Write the depth from which the sample was taken in Column 1. Weigh the sample to the nearest 0.01 g and record this weight in Column 2.

(2) Carefully place the sample into the sample cylinder. If two or more samples are to be measured simultaneously, weigh all the samples

before proceeding with the retorting. The sample should weigh between 75 and 120 g depending on the density of the rock and contained fluids. As a rule of thumb, the sample will fill about $7/_8$ of the sample cylinder. Screw on the lid handtight.

(3) Place the sample cylinders in the retort chambers making certain that each is properly seated so the discharge tube on the bottom of each cylinder projects below the heated chamber. This will avoid oil fires in the retort.

(4) Start the stopwatch. If more than two samples are done simultaneously, they should be placed by pairs in the retorts at 30-sec intervals.

(5) Liquids in the sample will now vaporise and be forced to discharge through the discharge tube into the condenser. Cooled by the water-cooled jacket, the liquid will then be collected in the graduated receiving tube below.

(6) Water and oil will collect simultaneously, thus the need for an emulsifier. Water will collect immediately to 3 min after the sample cylinder is placed in the retort chamber. Use a saturation data sheet such as the one shown in Fig. B–4 to record minute by minute the volume of water collected during the first 15 min of the measurement. The data sheet will clearly show at which time all the free water from the sample has been collected. This is shown in Fig. 1–7. The hydration of silts and clays will yield a small amount of water after approximately 15 min of retorting. Waters from hydration must not be used in the final calculations.

(7) Shut off the water passing through the condensers by closing with a clockwise motion the right-hand knob. This allows the heavier hydrocarbons to reach the receiving tubes.

(8) Drain the condenser jackets of water by pulling open the left-hand push-pull valve. At the end of a total of 40 min, remove the sample cylinder with a stiff wire or pliers and allow it to cool before handling. It may be placed under a water tap for rapid cooling. Record the amount of free water received in column 3. Measure the amount of oil collected and record it in column 4.

(9) In column 4 there are sub-columns for the temperature and gravity of the collected oil. The best but most improbable way to measure the recovered oil gravity on a drilling rig is with a refractometer. Since this piece of equipment is usually not available, an alcohol-water solution with a hydrometer is used. Fill a convenient transparent cylinder (glass case for the hydrometer itself) half full of water and put two or three droplets of oil on the surface of the water. Alternately, shake the cylinder and add small amounts of alcohol until beads of oil remain suspended in the solution, indicating that it is of the same gravity as the oil. Preferably

this is done with oil taken directly from the core as soon as possible. The temperature of the core is measured at the same time. This method will yield an accuracy within 2 points of the actual gravity when corrected from temperature with the API table shown in Fig. B–5, Appendix B. If retorted oil is used, the procedure is identical but the temperature of the retorted oil is measured as soon as the retorting is completed. This method is not accurate because retorting cracks the oil, and is not recommended.

(10) Clean the sample cylinders, especially the discharge tube, with a stiff wire brush. Water and solvent may be used but the cylinders must be dry before being used again. Clean the condensing tubes in the same manner. Wash and dry the receiving tubes. Caution: Solvents in hot retorts cause fires.

(b) Calculations.

(1) In Column 5 the Corrected Oil Gravity is recorded by using the table, Oil Gravity Temperature Correction, in Appendix B, Fig. B–5. Take the gravity determined in Column 4 and using the table, correct it for temperature.

(2) Column 6, the Corrected Oil Volume, is determined by using the graph (Fig. 1–6) which has been computed by Ruska titled Distilled Oil Volume Correction Factor. This graph has been constructed because the amount of oil measured in the receiving tubes is less than the actual amount contained in the sample. Residual oil is left through coking and adhesion to the grain surfaces in the sample. This correlation between the measured oil and the oil in the sample does not correct for the discrepancy between oil in the sample at surface conditions and oil in the sample at reservoir conditions.

(3) Column 7, Pore Volume in cubic centimeters, is determined by Formula 1, Fig. B–3. A second method may be used by first determining the bulk volume with the supplementary bulk-volume formula (Fig. B–3) followed by its corresponding pore-volume formula.

(4) Column 8, Percent of Water Saturation, is readily determined with Formula 2 which utilizes the entries in Columns 3 and 7.

(5) Column 9, Percent of Oil Saturation, is determined with Formula 3 which utilizes Columns 6 and 7. Columns 8 and 9 will not normally add up to 100% and the remaining percentage is assumed to be gas. If the two columns add up to more than 100% there is an error in the measurements or calculations. In tight carbonate rocks when the mercury-injection porosity measurement is used, this error will occur through no fault of the operator, since mercury will not penetrate the tiny capillary

openings. It is recommended that the air-injection method be used with such rocks to attain reasonable results.

6.2 Air Permeability Procedure

Air permeability is a measure of the ability of a viscous medium (air) to flow through openings between the pore spaces of a rock. It is important because fluid in a reservoir rock cannot be recovered unless there is freedom of movement from one pore space to another.

The core plug used for permeability measurement must be cut from the core close to the point where the saturation sample was taken. If it is not cut with a diamond drill it must be shaped in a form which can easily be measured and handled by the permeameter.

(a) Measurements. Consult the instruction manual for proper adjustment of the permeameter. The permeameter is factory calibrated but over a period of time it may become necessary to recalibrate the instrument. This is done by using a U tube to regulate the pressures in the instrument.

(1) Write the depth from which the sample was taken in Column 1 which corresponds with the samples for saturation and porosity. Measure its length to the nearest 0.01 cm with the vernier caliper, and record it in Column 10. If the ends are not perpendicular to the sides of the plug, take three measurements from different sides of the plug and use the average length. Measure the width of the plug and record it in Column 11.

(2) The plug, if it is to be used for porosity measurements (air-injection method), must now be weighed and its weight recorded in Column 30. Place the sample in the fluid extractor and remove it after three or four cycles of solvent washing. Place the sample in a vacuum oven or under heat lamps to dry. Put sample in the dessicator for cooling prior to making further measurements.

(3) Put the plug in the rubber sample holder; fit the metal sleeve around the holder and screw on the base plate. Be certain that the sample plug does not protrude from either end of the rubber sample holder.

(4) Fit the receptacle on the worm screw and tighten firmly but not tightly. The rubber sample holder will seal against the upper metal plate and the base plate.

(5) Turn the air-volume throttle clockwise until 0.25 atm is read on the large dial. Observe the ball in the large flowmeter. If the reading is below 2, continue with the procedure which follows. If greater than 2,

record the reading to the nearest tenth in Column 12. Record "large" in Column 13. Record the temperature measured with the thermometer on the permeameter in Column 14.

(6) Move the flowmeter selector arm opposite the word "medium" on the flowmeter selector case. Turn the throttle clockwise more carefully now to 0.50 atm. If the reading on the medium flowmeter is still below "2", the selector arm is moved to the "small" position and the throttle is turned until 1.0 atm is registered on the dial. Be cautious in turning the throttle not to force the floating ball in the flowmeter. The ball is easily damaged by such action caused by quick bursts of air. Write the reading to the nearest tenth in Column 12 and record which flowmeter is used in Column 13. Record the air temperature in Column 14.

(7) The same procedure is followed for vertical permeability measurements if required.

(b) Calculations. (1) Column 15, the Outflow Rate, is derived from a graph prepared by Ruska for the particular instrument in use. An example of such a chart is shown in Fig. 4–8. The actual chart to use is found in the Ruska Permeameter Manual. The instrument serial number is written on the graph. Notice that the paper is logarithmic in one direction and linear in the other. Take the flowmeter reading in Column 12 and graphically transpose it to the left-hand side of the graph which reads "Flowrator Reading in cm." Select the graphical curve which applies by referring to Column 13 and transpose the flowrator reading to the bottom of the chart titled "$\overline{Q}$ Outflow rate in cc/sec at mean Pressure $\overline{P}$." This is recorded in Column 15.

(2) Column 16, Air Viscosity, is determined by using the temperature recorded in Column 14 and applying it to the graph in Appendix B, Fig. B–6.

(3) Column 17, Cross Sectional Area of the sample plug, is determined by using Column 11 and one of the lines in Fig. B–7, depending on whether the plugs have circular or square cross-sections.

(4) Column 18, Darcys, has three equations which represent the figures obtained with either the large, medium, or small flowmeter tubes. "K" represents Darcys and the other letters symbolize Columns 10, 15, 16, and 17. Select the appropriate formula and work out the number of Darcys.

(5) Column 19, Millidarcys, is filled by dividing the figure in Column 18 by 1,000, which is the same as moving the decimal point three places to the left.

6.3   Mercury-injection Porosity Procedure

Three methods for measuring the porosity of a sample will be discussed, the first being the mercury-injection method. It is best to start
with this method to learn the basic operating principles of the Ruska
Porometer which is based on the Kobe porosimeter.

The mercury-injection method is accurate if careful technique is used
and the measurements made precisely. Errors result when air is trapped
in and around the sample. The sample cannot be reused or data checked
by a second run since the pore spaces have been filled with mercury.
Mercury will not penetrate the tiny capillary openings of some carbonate
rocks.

(a) Pycnometer Volume Calibration. Calibration is accomplished in
the beginning of an analysis and subsequently every four or five samples
thereafter. It consists of checking the volume of the pycnometer. Refer
to Fig. 4–6 for instrument terminology. On page 19 of the Ruska Porometer Manual there is information regarding the type and serial number of
the particular porometer being used. The volume of the pycnometer
vessel is given in cubic centimeters. Write this volume on a piece of mask-
ing tape and place it somewhere on the base of the porometer for future
reference. Also listed is a correction factor which should be written down
for reference. Both factors will vary with temperature and atmospheric
pressure changes.

(1) Remove the breech-locked lid from the pycnometer, after first
checking that the left and right-hand gauges register at or below zero and
the petcock valve on top of the lid is open.

(2) Turn the handwheel to the left (counter-clock-wise) until the
mercury level in the pycnometer vessel is at the inlet hole in the bottom
of the vessel. With a brush, clean the walls of the vessel of residual
mercury. If available, attach a vacuum extraction bottle to narrow tubing
and clean the vessel of any foreign matter . . . otherwise, clean the vessel
with soft tissue. *CAUTION: mercury is poisonous* if in contact with cuts
or breaks in the skin or if its vapors are inhaled.

(3) Check that the right-hand low-pressure gauge valve is closed.
Continue the left-hand motion with the handwheel until the mercury just
passes below the inlet hole in the bottom of the pycnometer vessel. Do
not let the mercury pass too far down the tube or air will be drawn into the
cylinder below and increase the margin of instrument correction factor
needed (para. (b)).

Slowly, take up the slack in the handwheel with a right-hand (clockwise) motion until the bead of mercury is as level as possible with the inlet hole at the bottom of the vessel. All motions of the handwheel while making measurements must be made to the right, taking up the slack in the handwheel-screw-piston assembly.

(4) Push the *volume-scale slide* forward until it is against the scale stop. Turn the handwheel slipping dial, reading the forwardmost right-slanting figures, until the decimal portion of the pycnometer volume given in the Ruska manual (now written on the porometer base) is at the handwheel dial index point. Be certain the handwheel has not moved.

If the pycnometer volume is 47.53 cc., for example, the volume-scale slide, when pushed forward to the stop, will read between 47 and 48 cc — roughly halfway between. The decimal portion of the figure is placed on the handwheel slipping dial (right-slanting numbers) at the index point.

(5) Turn the handwheel slowly to the right (clockwise) until zero on the handwheel slipping dial is at the index point. The volume-scale index should now read exactly as in the example above — 47 cc. If the index is not opposite the proper volume, adjust the screw under the volume scale slide until this is so.

(6) Put the breech-locked lid on the pycnometer, first checking the rubber seating gasket, and then tighten snugly to the right. The "F" on the lid surface should initially be on the right side of the instrument and almost facing the operator when tightened down.

(7) Turn the handwheel to the right, observing the volume scale as the index line approaches zero. Observe at the same time the petcock valve seat in the lid. As the mercury bead appears through the valve seat, back off a complete turn with the handwheel and proceed to the right more slowly, until the mercury is as nearly as possible level with the petcock valve seat.

(8) The volume scale should now be indexed at zero as should the handwheel slipping dial. If not, it is due to minor irregularities such as the pressure exerted to tighten the breech-locked lid, temperature or some similar factor.

DO NOT take off the breech-locked lid while the pycnometer is full.

Write down the volume actually measured. If the index for the handwheel dial is to the left of the zero, add the extra increments to the supposed volume. If the index for the handwheel dial is to the right of the zero, subtract the increments from the supposed volume.

(9) Back off to the left (counter-clockwise) with the handwheel several turns, remove the breech-locked lid, and continue turning the hand-

wheel to the left until the vessel is clear of mercury as before. Repeat the previous steps 4 through 8 in exactly the same manner three times to obtain an average volumetric reading. This then is the new volume of the pycnometer vessel. Check it by going through these same steps with the new measured volume.

(b) Instrument Correction Factor. Each porometer will have a different correction factor depending on the sensitivity of the individual instrument and any air trapped in the mercury system. Air is usually trapped in the gauges to some extent. If the correction factor is greater than 0.50 cc at 750 psi, the instrument must be purged of air. To do this, consult the Ruska Manual for the instrument. The instrument correction factor is determined in conjunction with the Pycnometer Volume Calibration (para. (a)). It is good practice to check this each time the instrument correction factor is checked.

(1) The pycnometer is full of mercury and the handwheel dial is indexed at zero, having just completed a pycnometer volume calibration. The handwheel is still tightly turned to the right (no slack). Close the petcock valve on the bead of mercury at the valve seat. Adjust the left-hand gauge dial to zero if not already on zero by removing the gauge face-plate and turning the adjusting screw. Recheck that the right-hand gauge valve is closed.

(2) Push the pore-volume scale forward against its stop. Turn the handwheel slowly to the right until exactly 750 psi is registered on the left-hand gauge. Hint: always stand in the same position when reading the gauge to limit the chance of distorted readings of the pressure measured on the gauge. All measurements must be performed in a similar manner.

(3) At 750 psi the index on the pore volume scale should be at zero. If not, adjust the screw under the scale. Note the numbers opposite the index of the left-slanting figures on the handwheel slipping dial. These should be less than 0.50 cc. If the porometer is a new piece of equipment or has not been used for some time, leave the pressure on the gauge for a few minutes to check for leaks at the petcock valve seat, the left-hand gauge, and the plunger-piston assembly. Suppose the number read on the handwheel slipping dial is 0.37 cc. This then is the instrument correction factor. It is standard practice that if the correction factor is greater than 0.50 cc the instrument must be purged of air. Consult the Ruska Porometer Manual for such a case.

(4) Turn the handwheel to the left until the gauge registers zero. Turn

the handwheel one full turn to the left "below" zero and open the petcock valve. After turning the handwheel down a few turns, the breech-locked lid may be opened.

(5) It is not necessary to go back to the starting steps for volume calibration but it is good practice in order to develop a systematic procedure. Repeat three times the entire procedure for Pycnometer Volume Calibration (para. 6.3(a)) and Instrument Correction Factor (para. 6.3(b)). The average reading of the measured correction factors is written on a convenient piece of tape and placed on the base of the porometer for handy reference.

(c) **Measurement.** After sizing a piece of the core for sampling with a hammer and knife, write the depth from which it was recovered in Column 1. Weigh the sample to the nearest 0.01 g and record the weight in Column 20. Examine the sample with the ultraviolet light for oil and mineral fluorescence.

Observe the sample under the microscope. Record the findings in Column 27. Extract the fluids from the samples for three or four cycles of solvent in the extractor or alternately dip the sample in toluene and acetone. Dry the sample in a vacuum oven or under heat lamps. Place the sample in a dessicator until cool.

(1) In Column 21, Weight of Porosity Sample Dry, record the weight of the dried and cooled sample. It is usually during the drying and cooling of the sample that the porometer is checked for pycnometer volume calibration and instrument correction factor.

(2) The volume-scale slide and the handwheel slipping dial provide direct readings in cubic centimeters of the bulk volume for a given sample in the pycnometer. Begin the procedure as in the pycnometer volume calibration in paragraph 6.3 (a)(1).

(3) When the pycnometer vessel is approximately half full of mercury, put in the sample to be analyzed being careful not to scrape the sample on the wall of the pycnometer. Place the breech-locked lid in place and tighten firmly. Leave the petcock valve open and turn the handwheel to the right. With experience one can judge approximately how many turns of the handwheel are required before it is necessary to slow the turning of the wheel while watching the petcock valve seat for the appearance of mercury. If too much mercury appears at the first try, back off one complete turn and bring the mercury to the valve seat again.

In Column 22, Bulk Volume, write the readings observed on the volume-scale slide (which are whole numbers) and follow them with the decimal reading taken opposite the index point on the handwheel slipping dial, using the right-slanting figures. The total reading, for example, might be 11.74 cc. A common error is to pick the wrong whole number on the volume-scale slide. Always use the number closer to zero opposite the index point for the whole number.

(4) Continuing from the above procedures, push forward the pore-volume scale to its stop. Set the instrument correction factor determined in paragraph 6.3(b) at the handwheel dial index, using the right-slanting figures. Again, be certain that the handwheel has not moved from its tight right-hand position. Close the petcock valve.

(5) Turn the handwheel slowly to the right until 750 psi is registered on the gauge. As mercury fills the pore space of the sample, the pressure on the gauge will decrease. Turn the handwheel until the pressure remains constant. Do not go over 750 psi since it would create an abnormal factor leading away from uniformity of sample measurement. Read the whole number on the pore volume scale, again using the number opposite the index which is closest to zero. Use the left-slanting figures opposite the handwheel dial index. Record the measured amount of pore volume in Column 23.

(6) Suitably mark the sample for later identification. Examine the sample under the microscope and complete the sample description in column 27.

(d) Calculations. The porosity calculations are very easy to make and several formulas are provided for reference on the worksheet. The grain volume and density (specific gravity) of the sample may vary a bit from the norm since the pore volume measured and recorded in Column 23 has been effective porosity calculations and not total porosity calculations.

6.4 Air-injection Porosity Procedure

The air-injection procedure is by far the most accurate field method available for measuring the effective porosity of rock samples.

In this system air is the compressing medium and the medium which already occupies the pore space. Thus the porosity is measured with a homogeneous medium of very low viscosity when compared to mercury. It is possible to analyze the samples a second and third time to recheck work since no mercury has been injected to destroy the porosity.

The permeability plug used to determine the permeability can be used for porosity measurement with this procedure, thus eliminating the necessity of cutting a separate plug for each measurement. Most important is the fact that if permeability and porosity measurements are made with the sample plug, the margin of discrepancy and error in the final results will be narrowed.

Three metal plugs are needed as explained in Chapter 4. By using the plugs to determine a sloped line on a graph, built-in errors in the porometer and the effect of atmospheric pressure are cancelled. It would be possible to make the sloped line on the graph paper by using the barometric pressure, but this would not permit the checking for any mechanical deviations in the instrument itself.

(a) Pycnometer Volume Calibration. Refer to Fig. 4–6 as a guide to terminology used to describe the procedure outlined below. Refer to page 19 of the Ruska Porometer Manual for data pertaining to the particular instrument being used. The pycnometer volume calibration is accomplished in exactly the same manner as described for the mercury-injection procedure, paragraph 6.3(a).

(b) Instrument Correction Factor. This is the first step which differs radically from what has been discussed in paragraph 6.3, the mercury-injection procedure.

Number and label a sheet of graph paper as shown in Fig. B–8. The paper should be large enough and defined with enough lines so that readings to the nearest cubic centimeter can be plotted. The object of the following exercise is to measure a given volume of air under compression in the pycnometer.

(1) Having completed the pycnometer volume calibration, turn the handwheel to the left (counter-clockwise) and empty the pycnometer of mercury. Be careful not to let the mercury fall too far below the inlet hole in the bottom of the pycnometer vessel.

(2) Open Valve No. 1 for the right-hand low-pressure gauge. Bring the bead of mercury to the inlet hole with a right-hand motion of the handwheel. Push the volume-scale slide against its stop. Set the cubic centimeter fraction of the pycnometer volume previously determined opposite the index line with the handwheel slipping dial, using the right-slanting figures.

(3) Turn the handwheel to the right until the volume-scale index is opposite 40 cc and the handwheel slipping dial is indexed at zero cc. The right-hand low-pressure gauge should be on zero. If not, remove the faceplate and adjust the calibration screw to put the dial pointer on zero.

(4) Place the breech-locked lid on the pycnometer, tighten as in the mercury injection procedure and close the petcock valve. Make certain

that the handwheel slipping dial has not loosened from its tight right-hand position. Sometimes it is advantageous to place the breech-locked lid on the pycnometer before the final turn made to the 40 cc mark in order not to disturb the tight right-hand position of the handwheel. Again, always do it in the same manner each time.

(5) With the petcock valve now closed and ignoring the left-hand high-pressure gauge, turn the handwheel to the right until 28 psi is indicated on the right-hand guage. Pause, then proceed slowly to 30 psi. Stand in a position behind the porometer from where all pressures on the gauge will be determined during calibration and sample measurements. It cannot be stressed strongly enough the importance for making measurements uniformly to narrow the margin of error in the analysis.

(6) If the porometer has not been used for some time or is a new piece of equipment, leave the pressure on the gauge for a few minutes to check for leaks at the petcock valve seat, the No. 1 valve or the plunger-piston assembly. The pressure should remain constant. The volume-scale index should point to approximately 10 or 11 cc. Always read the number nearest the zero on the volume-scale slide. Record this and the decimal portion on the handwheel slipping dial (right-slanting figures) opposite the index point in Column 28. This then is the number of cubic centimeters of air at 30 psi which was 40 cc of air at atmospheric pressure — zero pressure for this procedure.

(7) Referring to the graph in Fig. B–8, make a dot on the graph paper at the appropriate position on the Volume V_f axis. Since there is no solid object in the pycnometer, the *Grain Volume* V_g which equals zero is recorded in Column 29. In the particular example of Fig. B–8, the V_f is 10.60 cc. Release the air pressure slowly by loosening the petcock valve.

(8) To this point the air-injection procedure is similar to the Kobe method described in the Ruska Porometer Manual. The Kobe method relies, however, on a constant atmospheric pressure and assumes it, thus subjecting itself to unknown fluctuations in barometric pressure. Also, instrument and frictional distortions, where present, will not be detected and cannot be corrected for as with the air-injection procedure.

(9) Take the three metal plugs from their case. They are purposely of different sizes to create, with the volumetric measurement of each plug, a line (slope) which delineates, starting with the zero grain volume already measured in sub-paragraph (8) above, grain volume of increasingly larger plugs at the same pressure (30 psi). None of the metal plugs has porosity, so in effect the grain volume is the bulk volume.

An approximate measurement of the plugs might be 13, 14 and 15 cc on the V_f axis. These values will vary with each set of metal plugs but that is no matter as long as the same plugs are used during each suite of analyses. As accomplished in sub-paragraphs (5) and (6) with no plug in the pycnometer, repeat the procedure with each plug in the pycnometer vessel, starting with the smallest plug first. Place the figures obtained in Column 28.

(10) Do not turn the handwheel back to 40 cc after recording the V_f of each plug. Instead, open the petcock valve slowly, turn the handwheel to the right until a bead of mercury appears at the petcock valve seat. This produces on the volume-scale slide and the handwheel slipping dial (right-slanting figures) the volume of the plug in the pycnometer $- V_g$, grain volume. After measuring the volume of each plug, write the grain volume in Column 29 opposite the corresponding V_f in Column 28. Remember that for metal plugs the grain volume is identical to the bulk volume since there is no porosity. Approximate values for V_g corresponding to V_f values in sub-paragraph (9) are 3.0, 4.3, and 5.5 cc.

(11) Plot the figures recorded in Columns 28 and 29 on the prepared graph paper as shown in Fig. B–8. Draw a line between the plotted points. Label this line the Reference Volume Line or V_f Line.

Theoretically the line should be straight with all points falling on the line. Most likely, however, the plotted point for the medium-sized plug will be slightly off the desired straight line. This deviation is quite common with most Ruska porometers and it must be treated as an instrument idiosyncrasy. The amount of error created is at most only 0.002 cc.

To correct for this slight error, draw the line between the established points, not necessarily a straight line. If there are any variations in the straight line between other points on the graph, recheck the measurements and point plots. Any deviations in the line which create variances in porosity greater by 0.001% than would be otherwise determined with a straight line are reason enough with this system to warrant a mechanical check of the porometer.

(c) Reference Volume (V_f) and Bulk Volume Measurements. The sample which may also be the permeability sample is first weighed with its constituent fluids to the nearest 0.01 g, extracted of its fluids, dried in a vacuum oven or other suitable means, and cooled in the dessicator. Again, weigh the sample and write the appropriate figures in Columns 30 and 31.

(1) The instrument has been checked for pycnometer volume (para. 6.4(a)). Proceed so the instrument registers on the bulk volume scale just below or at 40 cc. Place the sample in the pycnometer; tighten the breech-locked lid; move the handwheel to the right until the handwheel

dial reads zero at the index and the bulk volume scale reads 40 cc at the index mark. Close the petcock valve.

(2) Turn the handwheel to the right . . . pause at 28 psi and then proceed slowly to 30 psi. Make certain that all gauge readings are read from the same position relative to the instrument in order to alleviate inaccuracies. Write in Column 32 the figures indicated on the volume scale (figures closest to zero) and handwheel dial (right-slanting figures) which make up the Sample Reference Volume, V_f.

(3) Loosen the petcock valve slowly; continue to turn the handwheel to the right until a bead of mercury appears at the valve seat. Turn back one complete turn if necessary and bring the mercury bead to the valve seat again. Record the figures indicated on the volume scale and handwheel dial in Column 33. This is the bulk volume.

(d) Grain Volume (V_g) and Calculations. The measurements necessary to calculate the porosity of the sample have been made. The graph is now used to determine the grain volume as shown in Fig. B–8 by the dashed line.

(1) Take the reference volume (V_f) recorded in Column 32, project it across the graph from the V_f axis and make a tick mark where it crosses the Reference Volume (V_f) line.

(2) Project the tick mark on the Reference Volume line down to the Grain Volume (V_g) axis. Record the indicated figure to two decimal places in Column 34.

(3) Subtract the grain volume from the bulk volume (Column 33 − Column 34) to establish the pore volume in Column 35.

(4) Column 36, 37 and 38 are self-explanatory with included formulas.

6.5 Grain Volume Porosity Measurement & Calculations

This method applies specifically to rock materials which have grains such as sands and sandstones and can be crushed with a mortar and pestal. The method is very difficult to apply to carbonate rocks due to their hardness and lack of individual grains.

(a) Measurements. The bulk volume of a shaped sample is measured with a Ruska porometer. It is not recommended to use the permeability plug which is cut with the diamond drill since the sample must be eventually destroyed to complete the measurements. It is very important that no grains or fluids are lost in making the following measurements.

(1) Write in Column 1 the depth from which the sample is taken. Weigh the sample to the nearest hundredth gram and record it in Column 39. Extract the fluids with the fluid extractor as described in the previous two methods or, if applicable, alternately dip the sample in toluene and acetone. Dry the sample with a vacuum oven or under a heat lamp and place in a desiccator to cool. Weigh sample after drying and record weight to nearest 0.01 g in Column 40.

(2) Obtain the bulk volume as in the mercury-injected method. Write the Bulk Volume in Column 41. If a portable pycnometer is used rather than the Ruska porometer, weigh the displaced mercury which is collected in a dish under the pycnometer. Multiply by the appropriate factor as regards operating temperature, from the chart "density of mercury" as found in Fig. B–10, Appendix B. This method also yields the bulk volume of the measured sample.

(3) Crush the sample to grain size using a mortar and pestal. Transfer the crushed grains to a liquid volumeter which contains a measured amount of toluene. Record the amount of fluid in Column 42. The increased volume of fluid observed, recorded in Column 43, is the volume of the toluene and the sand grains.

(b) **Calculations.** Subtract Column 42 from Column 43 and record the result in Column 44, the volume of sand grains. The pore volume equals the bulk volume minus the grain volume. Grain density equals the dry weight of the porosity sample divided by the grain volume. Total porosity equals 100 times the pore volume divided by the bulk volume. With the explained equations, Columns 45, 46, and 47 can be completed. The sample must be described before and after it has been crushed to benefit fully from this method.

7

Logging,
Testing, and
Estimating Reserves

Electric logs, self-potential logs, sonic log, formation-density log, and drill-stem tests add valuable data already gathered from the drilled and cored formation data about porosity, permeability and fluid saturation. In fact, advances in these fields have accelerated far more rapidly than the development of core analysis because of cost.

All the investigative procedures compliment each other. The drill-stem test offers the final conclusive test evidence of a formation's capability to produce hydrocarbons. With the drilling that delineates a prospective field, one can estimate the reserves of hydrocarbons that can be produced.

For several reasons, not all of which are scientific, attention is being more closely focused on the study of potential reservoirs with the aim of making larger hydrocarbon recoveries. Thus, core analysis is being used in conjunction with other methods to assess prospective discoveries.

The geologist interprets the electric log data as rock properties and fluids. Clean formations, especially sands and shales, are easily identified and quantitative interpretation is rather accurate. Where there is mixing of silt and/or "dirty" minerals and the formations more poorly sorted, quantitative interpretation is less accurate.

Carbonate rocks are even more difficult to interpret geologically from wireline logs because of their varied constituencies, primary and secondary porosities, mineralization and general physical make-up. For the ensuing discussion Schlumberger logging symbols are used (Fig. 7–1).

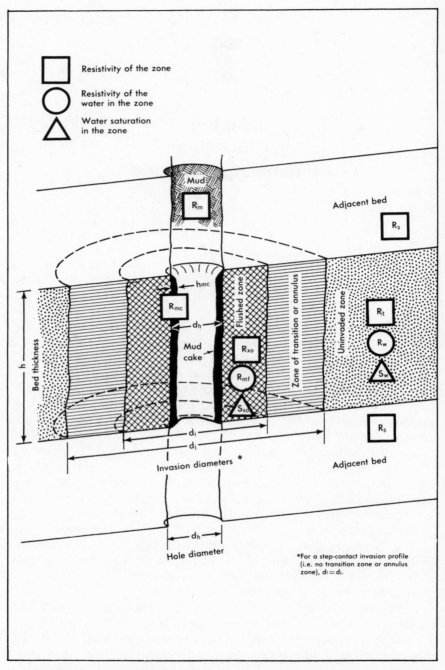

Figure 7–1 Symbols used in wireline log interpretation. Courtesy of Schlumberger.

7.1 Porosity, Permeability, Saturation

Clean sands, sandstones, oil, and gas are highly resistive and act as insulators in electrical-logging methods. Water, on the other hand, is conductive and receptive to electric logging.

In terms of resistivity, there is a constant value attained from the sandstone which is saturated with water called the Formation Resistivity Factor. This factor is increased or decreased through the incorporation of tortuousity and permeability into the otherwise pure problem of comparing the resistivity of a water-saturated sandstone with water alone.[60] Expressed another way, the Formation Resistivity Factor is a function of porosity, pore structure and pore distribution.

Archie found this relationship could be expressed by the equation:

$$R_o = FR_w \quad \text{or} \quad F = \frac{R_o}{R_w} \tag{7.1}$$

where: R_o = water-saturated sandstone resistivity
R_w = water resistivity
F = Formation Resistivity Factor.

R_o does not equal R_w as might be expected even though the fluids may be identical. The difference in R_o and R_w values is attributed to the tortuous paths the electric current must take around the insulator-like sand grains through the pore spaces to complete the electric circuit. This is the Formation Resistivity Factor. This may be expressed in a slightly different way with the empirical relationship:

$$F = \phi^{-m} \tag{7.2}$$

where: m = variable cementation factor
ϕ = porosity

Figs. 7-2 and 7-3 exhibit the relationship between the resistivity factor, porosity, and permeability. Notice the spread of values in formation factors depending on the cementation and compaction of the formations. These empirical values very often differ, even locally, from well to well in the same or like formations.

Fig. 7-4 shows the relationship of formation factors to porosity.

$F = \dfrac{0.81}{\phi^2}$ applies to sands; $F = \dfrac{1}{\phi^2}$ applies to compacted formations such as

chalk; $F = \dfrac{1}{\phi^{2.2-2.5}}$ applies to compacted, oolicast-type formations.

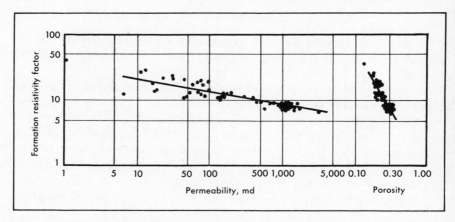

Figure 7-2 Relationship of porosity and permeability to formation resistivity factor. ARCHIE, Courtesy of SPE or AIME.

Where the value of "m" cannot be determined, the Exxon Co. formula can be applied:

$$F = 0.62 \, \phi^{-2.15} \tag{7.3}$$

This equation works well when applied to sucrosic types of rocks.

The formation resistivity factor may be applied to the determination of saturation values where the formations are substantially clean. The empirical relationship is known as the Archie equation.

$$S_w = \left(\frac{R_o}{R_t}\right)^{1/n} \tag{7.4}$$

where: S_w = Water saturation

R_t = Resistivity of formation containing hydrocarbons and formation water.

R_o = Resistivity of formation when 100% saturated with formation water.

It has been found that values of n range between 1.7 and 2.2 depending on the rock. Since n = 2 gives a reasonable approximation, the following formula results when substituting from Formula 1 the equivalent of R_o.

$$S_w = \left(\frac{R_o}{R_t}\right)^{1/2} = \left(\frac{F R_w}{R_t}\right)^{1/2} \tag{7.5}$$

As the formations become dirtier with shale and silty material, the reliability of the values used in the formulas become suspect and the process

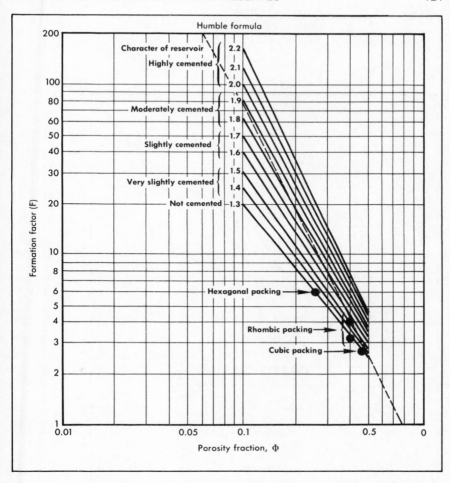

Figure 7–3 Formation factor versus formation porosity. From Oil Reservoir Engineering by Sylvain J. Pirson. Copyright 1958, McGraw-Hill Book Company.

of developing reasonable data becomes increasingly more detailed and complex.

Resistivities vary with the type and amount of fluid content. The amount of fluid is most often dependent on grain size and compaction. Potential reservoir formations may be classified as soft, intermediate and hard.

Soft formations are poorly sorted and poorly consolidated sand-shale sequences. The porosity of the sand is above 20% and the resistivity

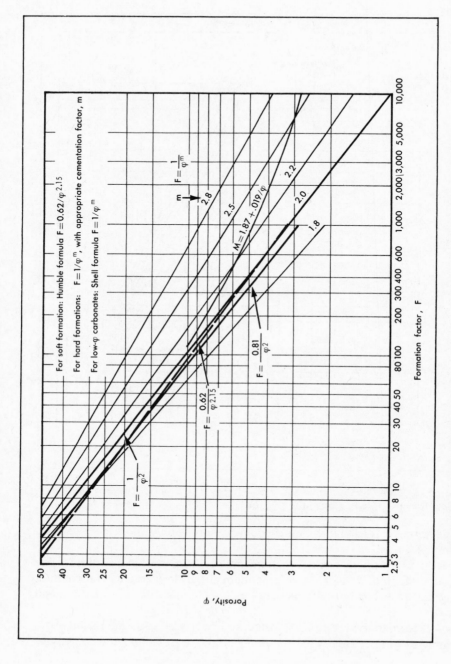

Figure 7-4 Formation factor versus porosity. Courtesy of Schlumberger, Log Interpretation Charts.

ranges from 0.3 ohm-m for saltwater sands to several ohm-meters for oil-saturated sands.

Intermediate formations are made up of fairly well-consolidated sandstones or limestones or dolomites. The porosity varies between 15–20% and the resistivities range from 1 to 100 ohm-m depending on the amount of interbedding of shales and/or densely packed (tight) rocks.

Hard formations are usually limestones and dolomites but can in certain provinces include well-cemented tight sandstones and orthoquartzites. The porosity is generally below 15% in hard formations and often the porous-permeable zones are made up of fissures, fractures and vugs. The resistivity ranges from 2 or 3 ohm-m to several hundred ohmmeter.

Completely impervious formations such as anhydrite, halite and to some extent coal, have extremely high resistivities due to their low fluid content.

Where hydrocarbon-bearing carbonates are considered by themselves, they may be graded with regard to resistivity and porosity. Fig. 7–5 shows

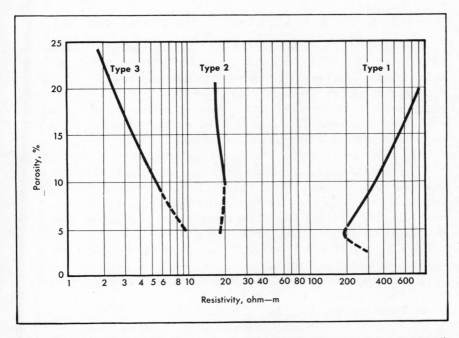

Figure 7–5 Porosity and resistivity relationships in carbonates which contain oil. After ARCHIE, Courtesy of AAPG.

the character of three grades of carbonates. The first, Type 1, initially drops in resistivity as the porosity increases because all the porosity is primary and water-saturated. Above the 5% porosity level, the resistivity increases, as secondary porosity plays an increasing role through the introduction of channeling and vugs. Increased secondary porosity causes a decrease in capillary water and an increase in resistivity. Oil is likely to be in place in this type of rock where the resistivity is high.

Type 2 carbonates depict uniformity which is rare to this classification. Where it occurs as shown, it means that water saturation and capillary forces remain proportionally balanced with respect to the formation. The rock is stable with only primary porosity and permeability — a characteristic of uniformly deposited shallow-water sediments.

Type 3 carbonates on the other hand, have a constant (relatively speaking) capillary water saturation with increasing porosity. The oil is usually found in the section with the highest porosity and the lowest resistivity. The porosity and permeability are primary and the sedimentary deposition occurs in deep marine water.

7.2   The Self (SPONTANEOUS) — Potential (SP) Log

The SP log is a measurement of the electrical potential energy in the mud around the sonde as compared with a reference electrode grounded at the surface. It is the most widely understood and easily "read" log and one which is too often wrongly used as a basis for making profound statements about the qualities of formations down the hole. There is no definite relationship between the magnitude of its curves and the porosity-permeability values of the rock formations.

The mud-logging drilling-rate curve approximates the curves of the SP Log. The SP curves are usually more useful in correlating formations between wells, to scan for permeable beds, locate formation boundaries, and obtain values of the formation water resistivity. The SP curve is a guideline indicating the points and zones where in-depth study must be made with the other wireline tools.

The factors which influence the character of the SP curve are:
1) The thickness of the bed.
2) The resistivities of the bed, adjacent formations and the mud.
3) Hole diameter.
4) Depth of mud invasion.

The amplitude of the curve varies as the value of the total electromotive force. In soft formations the SP curve defines the formation boundaries with a good deal of accuracy. In hard formations there are less distinct slopes and less discernible boundaries.

Other electric logs are used in conjunction with the SP log, and in oil-based muds it is necessary to substitute an induction log, gamma-ray log, or neutron log to find the permeable beds. There is a method to calculate the R_w (Formation Water Resistivity) from the SP log. Where R_w values are less than 0.1 they can be considered reasonably accurate. Where larger, they are not so accurate but may be the only data available.

The best measurement for R_w is one which uses a sample of formation water. If this is not practical, the R_w from nearby wells drilling the same formation may be available.

7.3 Radioactivity and Acoustic Logging

Radioactive logs are especially valuable for formation identification. Acoustic logs are valuable for evaluating the amount of porosity.

Saltwater and oil muds often make formation delineation very difficult with electrical logging tools, so radioactivity logging tools are sometimes mandatory, especially in carbonate formation investigations.

A thorough treatment of the subject may be found in Schlumberger's *Log Interpretation Principles*. The object of this section is for familiarization—to make more detailed examinations of the subject meaningful and objective.

There is an unusually high natural radioactive concentration in shales and clays because of their ability to absorb radioactive igneous minerals such as micas and feldspars from decomposed igneous rocks. The isotope K_{40} is an important feature of their make-up.

Sandstones, being purer mineralogically, are radioactive only insofar as they have "dirty" minerals in the matrix cement.

Primary limestones, having developed from marine organisms mostly, are low in radioactive elements.

Dolomite, secondary limestone, and accretionary limestones may exhibit some radioactivity through the introduction of dirty sediments and fluid concentrations from radioactive sources. As can be seen in Fig. 7–6, organic marine shales have the highest radioactive concentration.

The gamma ray log looks much like the SP log used in electrical logging. It is used to differentiate between shales and nonshales (sandstones and limestones). The gamma ray curve shows medium to high values for shale, while the neutron curve records low values for shale (siltstone, clay,

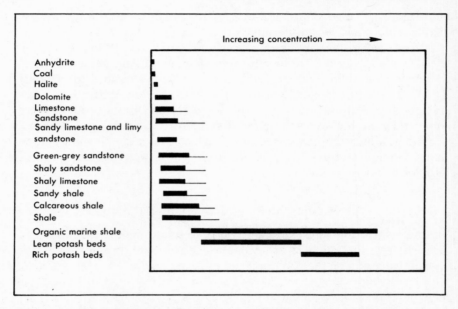

Figure 7-6 Radioactivity concentrations with respect to various sedimentary rocks. After Russell, Courtesy of AAPG.

marl or slate). One exception to these general values is a clay, rich in kaolin (probably from orthoclase feldspar), which registers quite low radioactivity values. Limestones, chert, halite, anhydrite, chalk, sandstone and dolomite all give similar but not identical responses—low gamma ray readings and high neutron readings.

Oil, once thought to necessarily be high in radioactivity, is actually low in this respect. This is yet another hint that it generates from marine organisms in closed deepwater basins. Fig. 7-7 shows distribution of radioactivity with respect to sediment types.

The effect of shading in shale-free limestone and dolomite appears to indicate the intensity of radioactivity in the samples as shown in the following abstract comparison.

	Average radioactivity equivalents per gram
Light grey to white	3.1
Medium shade	4.1
Dark to black	6.1

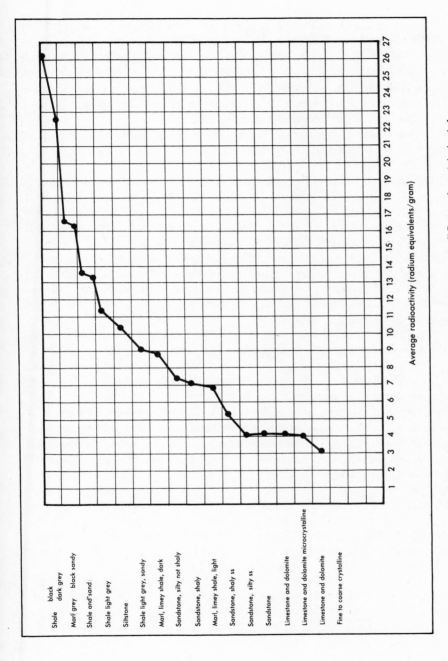

Figure 7–7 Radioactivity levels of sedimentary rocks.[68] Drawn from tabular information by Russell, Geophysics.

The gamma ray log can be used in cased holes where it can substitute for the SP log. It reflects the shale content of formations. The neutron log on the other hand is a measurement of the relative amount of fluid surrounding the sonde in the borehole. Basically, it delineates porous formations and this information is used to determine porosity. This is accomplished by measuring the hydrogen content, whether the fluid be tar, liquified gas, petroleum, fresh or saltwater.

An increase in hydrogen content indicates an increased amount of fluid and pore space. This increase is reflected in neutron logging by a decrease in the neutron count rate.

Because they are so complimentary, the gamma ray and neutron logs are run together. The gamma ray curve differentiates between the *possibly* porous (cherts may be fractured) and the nonporous (meaning impermeable) rocks; the neutron curve shows which rocks are porous, with attendant permeability. (Most discussions on wireline logs define shale as a nonporous rock as compared to porous reservoir rocks, a point worth remembering in geological discussions with petrophysicists and the like).

With the "shale line" defined as a reference by the gamma ray curve, the neutron curve can be regarded as a shale-plus-liquid indicator, which adds up to usable porosity.

There are two variations of neutron logging tools — a GNT neutron logging tool and an SNP logging tool. The first can be run in cased holes and the logs are registered in API units. The second (SNP) must be run in the open hole and gives gamma ray information in API units but neutron information in percent porosity.

These radioactivity logs cannot, by themselves, be used to differentiate between water and oil since both have such similar hydrogen content. A resistivity log must be used in conjunction with the radioactivity logs to see the difference. Fig. 7–8 shows typical gamma ray and neutron responses to various formations.

Where a porous zone is filled with gas (methane, ethane and sometimes propane), the neutron curve will ordinarily register very high intensities because of the lack of hydrogen in liquid form. It can be distinguished from high intensities registered with shale through comparison again with the gamma ray curve. It will clearly show a gas-liquid contact and is thus useful in exploitation drilling of a particular field.

The formation density log is also a porosity-logging tool. Using a gamma ray source and the electron-scattering principle (caused by bombardment with gamma ray particles) it counts the scattered gamma rays which reach the detector. This indicates the formation density by its

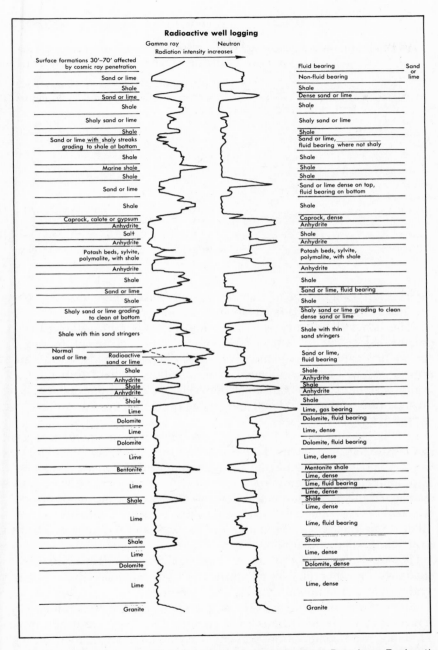

Figure 7–8 Comparison of typical formation responses. From Petroleum Exploration Handbook, G. B. Moody, Copyright 1961 used with permission.

TABLE 7–1. COMPARISON OF MATERIALS, VELOCITIES AND TRANSIT TIMES
(Courtesy of Schlumberger)

Material	Sonic velocity, V (ft/sec)	Transit time t (= 10 /V, in usec/ft)
Air (S.T.P.)	1,088	919
Methane (S.T.P.)	1,417	706
Oil	4,300	232
Water (mud)	5,000–5,300	200
Neoprene (typical)	5,300	189
Shales	6,000–16,000	167–62.5
Rock salt	15,000	66.7
Sandstones	18,000–21,000	55.6–47.6
Anhydrite	20,000	50
Limestones	up to 23,000	47.6–43.5
Dolomite	23,000–26,000	43.5–38.5

direct relationship to the measured electron density (number of electrons per cubic centimeter). It is called an FDC log.

The acoustic velocity log (sonic log) measures porosity when it is greater than 5% and less than 30%. It does not delineate soft formations very well since it is too responsive to the contained fluids. Quantitative porosity values can be obtained in the medium to hard formations. It is the best toll for distinguishing between salt and anhydrite. The tool must be run in an uncased hole but can be run in any type of drilling fluid except air.

Table 7–1 and Fig. 7–9 comprise the reference data guidelines used in acoustic logging. The mud invasion of a formation affects the measured velocity and must be taken into consideration when calculating velocities. The numbers and the graph are compiled for moderately invaded, well compacted, clean formations. Where oil and gas-bearing sands are considered, it is recommended that the resulting measured porosity as established from Figure 7–9 be multiplied by 0.85 and 0.70 respectively to establish the real porosity.[71]

The sonic log does not measure permeability as might be expected. Shaliness or other dirt in the formation will influence the measured transit time and ultimately the calculated porosity.

Since the neutron, sonic, and formation density logs react differently and independently to various rock compositions they are useful in determining rock composition porosity, and mineral breakdown through crossplotting of information gained from the combination of logs.

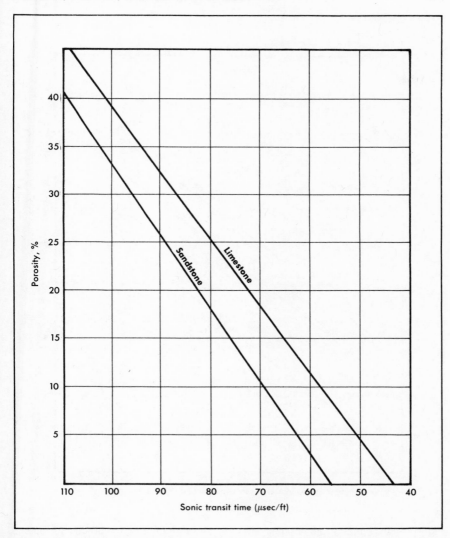

Figure 7-9 Relationship of porosity and transit times for limestones and sandstones.

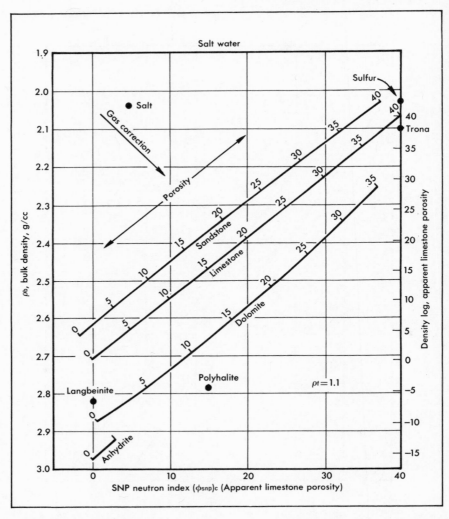

Figure 7-10 Porosity and lithlolgy determination using the formation density log and the sidewall neutron porosity log (SNP). Courtesy of Schlumberger.

7.4   Drill-stem Tests

Drill-stem testing was first introduced by Halliburton in 1926. It has become the "real test" of whether a formation will be a producer or not. On the results of this test, after considering all the accumulated data, the

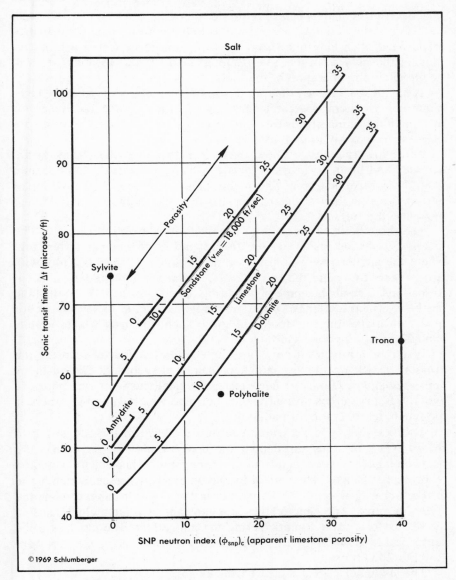

Figure 7-11 Porosity and lithology determination from sonic log and sidewall neutron porosity log (SNP) may also be used for GNT, F, G or H neutron logs. Courtesy of Schlumberger.

final decision is made whether or not to run casing and complete the well.

The basic function of the test tool is to temporarily relieve the hydrostatic head of drilling mud from the formation to be tested without removing the mud from the hole. Reference to Figs. 7–12 and 7–13 will help explain the operation of the tools.

It has become more common to use the dual closed-in pressure technique so that there is greater certainty of gathering accurate reservoir information from a single test. The terms "closed in" and "shut in" are used synonymously in this discussion.

After lowering the made-up tools into the hole, the packer(s) are actuated and form a bridge or blockage which separates the mud column from the zone to be tested. The anchor pipe, or tail pipe, is the section of perforated pipe below the packer through which the formation fluid passes to the surface.

The three basic types of tool assemblies are shown in Fig. 7–12. The first is for testing inside casing; the second is an open hole test tool where the anchor pipe sits firmly on the bottom; the third is a tool with two packers to isolate the test formation from the immediately adjacent formations. The anchor pipe is still firmly on the bottom. When the hole has been drilled too far below an interesting formation to set the anchor pipe with a straddle test tool, a sidewall anchor is used which permits enough force to operate the tool.

Water or nitrogen cushions are quite often applied to decrease the pressure shock on the formation, packer, and recorder. The cushions are especially required for deep tests. By using cushions one hopes to avoid pressure surges around the anchor pipe which will plug if there is too much debris in the formation fluid.

The bourdon tube pressure recorder is placed in the test string to record well pressures throughout the test period. The recordings are made on a black-coated metal chart by a movable stylus which is actuated by pressure changes. The mud of formation pressure presses a membrane in the tool wall which affects the fluid in the bourdon tube, causing the stylus to move. The chart moves at a constant rate like a clock causing the stylus to register a mark during the entire test. The clocks can be set for 3, 12, 24, 48 or 72 hr. The pressure ratings of the gauge vary from 1,500 to 20,000 psi.

The six major pressure measurements are:

(1) IHP; initial hydrostatic pressure
(2) IFP; initial flow pressure
(3) ISI; initial shut-in pressure (initial closed-in pressure)
(4) FSI; final shut-in pressure (final closed-in pressure)

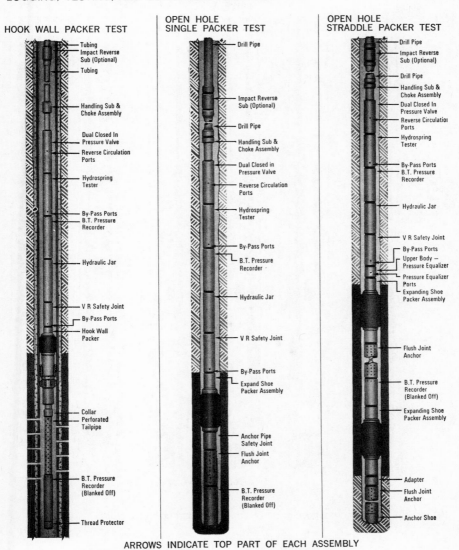

HOOK WALL PACKER TEST

- Tubing
- Impact Reverse Sub (Optional)
- Tubing
- Handling Sub & Choke Assembly
- Dual Closed In Pressure Valve
- Reverse Circulation Ports
- Hydrospring Tester
- By-Pass Ports
- B.T. Pressure Recorder
- Hydraulic Jar
- V R Safety Joint
- By-Pass Ports
- Hook Wall Packer
- Collar
- Perforated Tailpipe
- B.T. Pressure Recorder (Blanked Off)
- Thread Protector

OPEN HOLE
SINGLE PACKER TEST

- Drill Pipe
- Impact Reverse Sub (Optional)
- Drill Pipe
- Handling Sub & Choke Assembly
- Dual Closed in Pressure Valve
- Reverse Circulation Ports
- Hydrospring Tester
- By-Pass Ports
- B.T. Pressure Recorder
- Hydraulic Jar
- V R Safety Joint
- By-Pass Ports
- Expand Shoe Packer Assembly
- Anchor Pipe Safety Joint
- Flush Joint Anchor
- B.T. Pressure Recorder (Blanked Off)

OPEN HOLE
STRADDLE PACKER TEST

- Drill Pipe
- Impact Reverse Sub (Optional)
- Drill Pipe
- Handling Sub & Choke Assembly
- Dual Closed In Pressure Valve
- Reverse Circulation Ports
- Hydrospring Tester
- By-Pass Ports
- B.T. Pressure Recorder
- Hydraulic Jar
- V R Safety Joint
- By-Pass Ports
- Upper Body — Pressure Equalizer
- Pressure Equalizer Ports
- Expanding Shoe Packer Assembly
- Flush Joint Anchor
- B.T. Pressure Recorder (Blanked Off)
- Expanding Shoe Packer Assembly
- Adapter
- Flush Joint Anchor
- Anchor Shoe

ARROWS INDICATE TOP PART OF EACH ASSEMBLY

Figure 7–12　Three basic tool assemblies. (Courtesy Halliburton Co.)

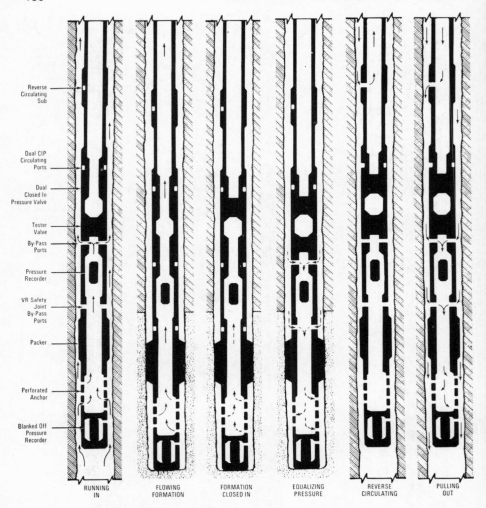

Reverse
Circulating
Sub

Dual CIP
Circulating
Ports

Dual
Closed In
Pressure Valve

Tester
Valve
By-Pass
Ports

Pressure
Recorder

VR Safety
Joint
By-Pass
Ports

Packer

Perforated
Anchor

Blanked Off
Pressure
Recorder

RUNNING
IN

FLOWING
FORMATION

FORMATION
CLOSED IN

EQUALIZING
PRESSURE

REVERSE
CIRCULATING

PULLING
OUT

Figure 7–13   Fluid-passage diagram. (Courtesy Halliburton Co.)

    (5) FFP; final flow pressure

    (6) FHP; final hydrostatic pressure

    Fig. 7–14 shows a test where there is only one flow period. Fig. 7–15 shows a test where there are two flow periods. The reason for two flow periods is explained later. Between points A and B the hydrostatic pressure increases as the tools are lowered in the hole. At point B the packer

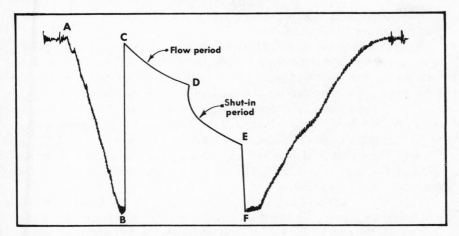

Figure 7–14   Typical drill-stem test chart.

is set and the formation is relieved of the hydrostatic pressure as the tester valve is opened and the chart is scribed to point C.

As the formation produces fluid a hydrostatic head is created as shown from C to D. This is called a "flow period".

At point D the shut-in pressure valve is closed, stopping flow of the

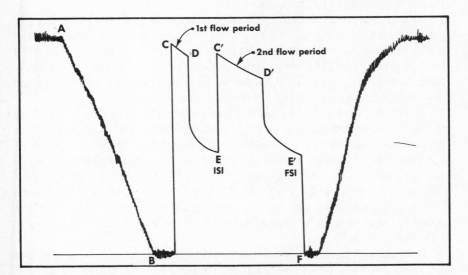

Figure 7–15   DST chart where dual closed-in pressure method is used.

formation fluids. Between D and E the formation pressure builds to near static pressure in the formation.

The shut-in pressure (FSI) is always less than the actual formation static pressure unless the shut-in period is allowed to extend for enough time so the line becomes parallel to the pressure reference marks. The time required to accomplish this cannot be readily ascertained at the surface so the shut-in time ultimately used is a matter of experience with the formation in the area. Nevertheless, a formation static pressure can be extrapolated from the pressure buildup of the curve.

At E the bypass valve is opened and the gauge again registers hydrostatic pressure as shown at F. The packer is then unseated and the tools are retrieved to the surface.

Often two recorders are used; one is made up so that it is in the flow-stream in the pipe above the packer and the other is set at the bottom of the anchor pipe and blanked off from the internal diameter of the pipe. The two recorders are used where there are perforation-plugging possibilities as the tester valve is opened.

In a good test, the two recorder charts will be identical except for a slight difference caused by their respective positions with regard to hydrostatic pressure.

In a test where plugging occurs, the blanked off recorder, which senses the formation pressure from outside the pipe, will continue to record the pressure present at the formation wall.

The end of a successful DST, where there is promising evidence of oil or gas (especially gas), can be a worrying time for a mudlogger and the driller. It used to happen that a cutback in the mud weight was ordered based on some "careful scrutinizing" of the DST chart final shut-in pressure. The watering back of the mud was ordered to increase drilling speed and to avoid the possibility of formation breakdown.

This was extremely dangerous since, as explained previously, the static formation pressure is rarely registered on the pressure chart. This type of situation is now avoided by the almost universal acceptance of the dual closed-in (shut-in) pressure method of testing. It utilizes a technique in which the shut-in pressure buildup (D to E) is recorded before the real flow period (C' to D') is measured. This allows a quick, more accurate measurement which can be extrapolated to the static formation pressure, than can be achieved after the disturbance caused by a long flow period (Fig. 7-15).

Loggers still have a worry however. The inertia set up by the test and the eventual drawing up of the tools can allow a dangerous amount of gas to bleed into the hole. It is frankly best not to water back mud at all based on a DST chart until the hole has been circulated several times. There is no need to mention the case of a good "show" and an obvious requirement to weight up the mud for this action is towards the positive side of safety.

The IHP and FHP must be equal for the test to be considered realistic and valid. The FFP should equal the hydrostatic head of fluids recovered in the drill pipe.

Quantitative analyses can be made from the scribed lines on the recovered charts. A special "chart reader" instrument is necessary to develop the precise information from the charts.

The reliability of drill-stem test records to show the characteristics of a potential reservoir formation is dependent on the quality of the test itself and the measurements of fluid recoveries. The experienced tester can tell a lot about how a test is progressing through the use of the "bubble bucket." This is simply a pail of water with a hose from the test-

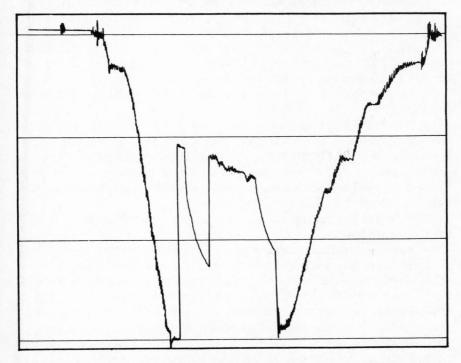

Figure 7-16 Example of chart where gauge is not zeroed before the test to the scribed line. Anchor pipe plugging is shown during second flow period. Redrawn after Murphy, Courtesy of Halliburton.

ing control head through which air in the drill pipe is allowed to escape and bubble in the bucket. The length of time required to dissipate the supercharge from the formation during the first flow period is best determined through use of the bubble bucket.

A slow bubble rate means a rather long (30 min or so) initial flow period is required; a fast bubble rate means a short flow period is in order.

The interpretation of the DST chart characteristics is developed through accurate plotting of the data scribed on the chart and a thorough understanding of the tools and formation behavior under test. It is beneficial to point out some of the curiousities that occur while making a DST by showing some chart patterns. See Figs. 7-16, 7-17, 7-18, and 7-19.

The DST is a reservoir-evaluation method as well as an exploration method. Because the chart will show the "skin condition" (the helping and hindering factors towards formation production) of a well, the recovered test fluids will not alone be relied on for determining potential recovery. Figs. 7-20 through 7-27 show some of the conditions which are recorded during drill stem tests. Fig. 7-28 shows how the chart is used to extrapolate the pressure information.

Table 7-2 gives examples of well-performance gain after completion treatment. The treatment is directed after analyzing the DST charts and apparent limitations of the formation to produce efficiently.[89]

In normal situations the following points should be heeded when making a DST.

1. Watch the "bubble bucket" for changes in the size and amount of bubbles which indicate what is happening during the flow periods.

2. The recovered liquid in the drill pipe should be carefully measured, noting where the liquid first appears, and where it is and isn't contaminated with mud and or saltwater. Make density measurements of the fluids described.

3. Measure the gas flow at several equally spaced intervals during the flow periods.

4. Reduce the pressure drawdown on gas tests by using chokes.

5. The first flow period should be at least 5 min long followed by an initial shut-in period of 30 min or more. The length of time for the second flow period is decided from experience in the area, the character of the formation geologically, and the type of show indicated on the mud-logging equipment. The final shut-in period must equal in time the second flow period.

6. Evaluate the accuracy of the pressure gauges.

7. Calculate the hydrostatic pressure of the recovered fluids and compare it with the pressure recorded for the final flow period.

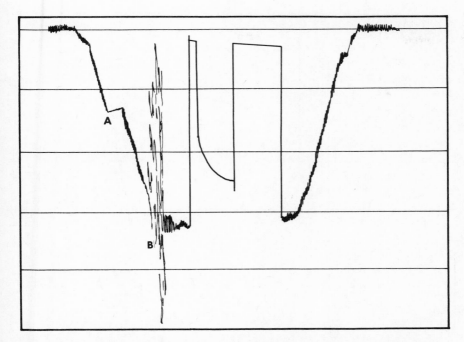

Figure 7-17 Leaking drillpipe at "A" evident from decrease in hydrostatic pressure when it should remain steady. A tight hole is shown at "B" by the wild fluctuations in pressure registered by the stylus. Redrawn after Murphy, Courtesy of Halliburton Co.

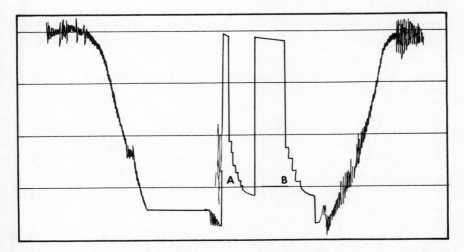

Figure 7-18 Example of a "stair stepping" gauge as shown at A and B. This is caused by mechanical sticking of the chart. Redrawn after Murphy, Courtesy of Halliburton Co.

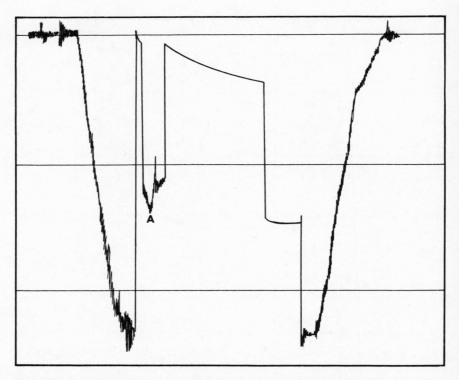

Figure 7–19 Leaking dual closed-in pressure at "A." Redrawn after Murphy, Courtesy of Halliburton Co.

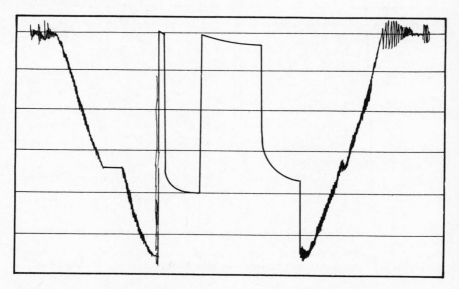

Figure 7–20 Supercharged initial build-up curve during shut-in period. This is caused by surges in the formation or by an excessive hydrostatic head prior to setting the packer. The curve is difficult to distinguish from depleting reservoir condition without extrapolation. Redrawn after Murphy, Courtesy of Halliburton, Co.

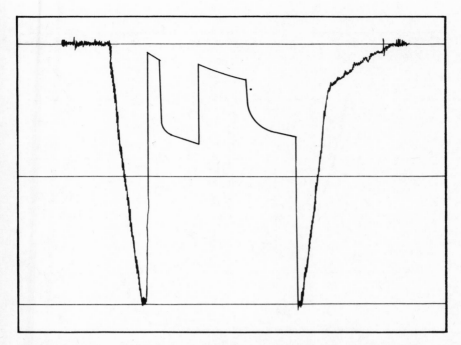

Figure 7–21 A barrier which is indicated in the buildup curves during the shut-in periods. A barrier is best seen on an extrapolation but on the chart is recognized by two radii—the first is short and the second is long giving the appearance of a flattening curve. Redrawn after Murphy Courtesy of Halliburton Co.

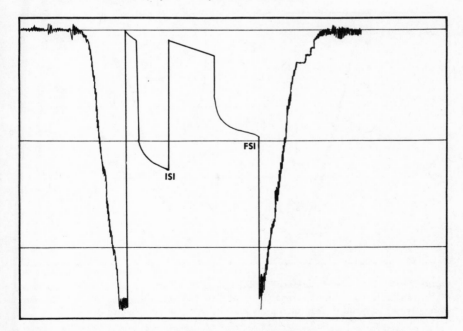

Figure 7–22 Example of a depleting liquid reservoir. Notice the slope differences between the first and second flow periods. From the ISI and FSI it is seen they will not extrapolate to the same point. Redrawn after Murphy, Courtesy of Halliburton Co.

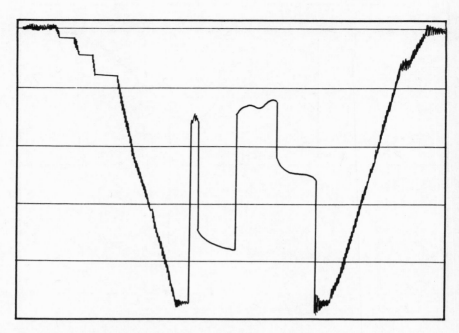

Figure 7–23 Example of a depleting gas reservoir. There is no flow into the anchor pipe as shown by the two flow periods. The initial build-up curve and ISI vary considerably from the second shut-in period and FSI indicating a decrease in pressure. Redrawn after Murphy, Courtesy of Halliburton Co.

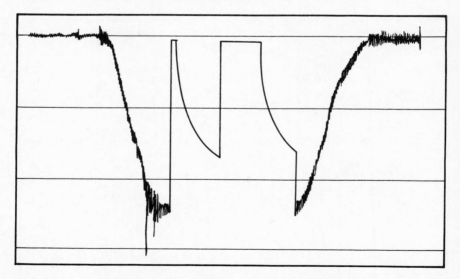

Figure 7–24 Example of low permeability (low productivity) as shown by flat horizontal flow periods. Uniform high reservoir pressure is indicated by slopes and final positions of shut-in periods. Redrawn after Murphy, Courtesy of Halliburton Co.

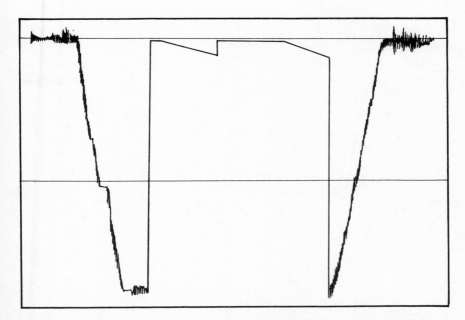

Figure 7–25 Example of low permeability (low productivity) as shown by flat horizontal flow periods. Very low reservoir pressure is indicated by the flat shallow shut-in periods. Redrawn after Murphy, Courtesy of Halliburton Co.

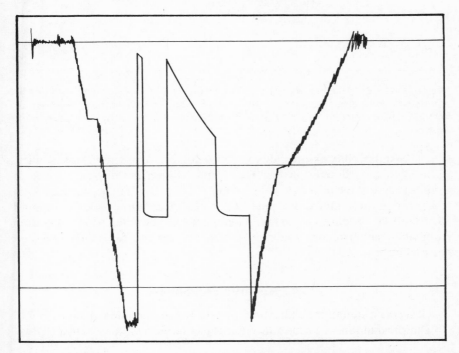

Figure 7–26 Example of high permeability (high productivity) as indicated by steeply sloping flow-period curves. Notice the second flow period curve is steeper than that of the first. Very high formation damage is indicated by the barrier evident from the two shut-in curves which quickly become horizontal. Redrawn from Murphy, Courtesy of Halliburton Co.

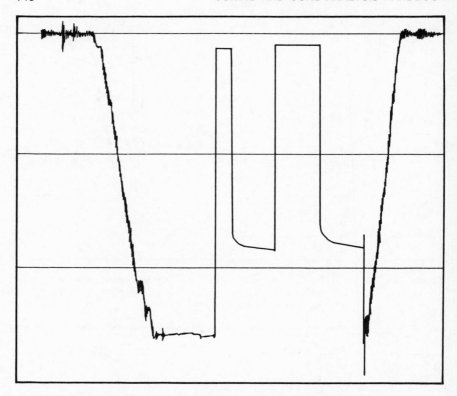

Figure 7–27 Example of low permeability (low productivity) as shown by the horizontal flow-period traces. Relatively high formation pressure and high damage are indicated by the character of the shut-in periods. Redrawn from Murphy Courtesy of Halliburton Co.

Transmissibility is an index of the effective formation permeability. The average effective permeability is the ability of the pore spaces to conduct fluids when they are less than 100% filled. The damage ratio is the productivity index of inherent formation characteristics compared to the DST productivity index. The approximate radius of investigation indicates the distance from the well bore that the formation was disturbed during a test.

7.5   Reserve Estimation

Reserve estimation is the study of reservoir phenomena as they relate to the production of crude oil, natural-gas liquid, and natural-gas. Most

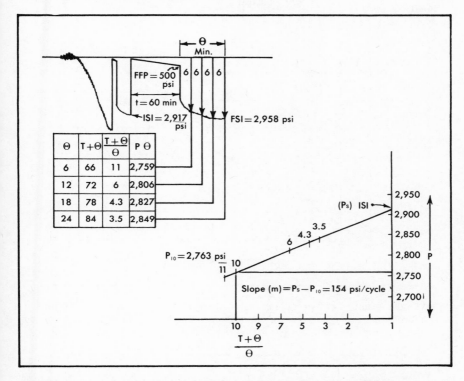

Figure 7–28 Method for plotting extrapolated chart slope from the DST gauge chart. Redrawn from VanPoolen and Bateman, reprint, courtesy of Halliburton Co.

oil pools have several energy sources none of which will continuously predominate through the life of the pool.

In its simplest form the recoverable reservoir oil volume is shown by the following formula.

$$\text{Recoverable stock-tank oil (bbl)} = (A)(t)(F)$$

where: A = area of producing formation, acres
         t = formation thickness, ft
         F = recoverable oil, bbl/acre-ft

When the porosity, permeability, and interstitial water saturation have been determined along with their relationship to the oil, a slightly more detailed formula can be utilized.

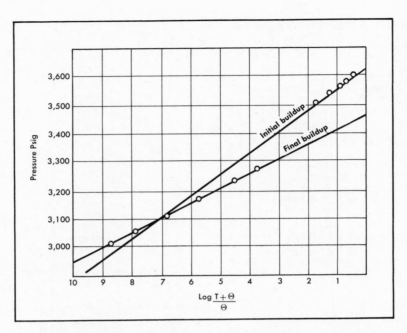

Figure 7–29 Example of extrapolated curves from initial and final shut-in periods. A false impression is gained from the supercharge buildup of the first shut-in period indicating the first flow period was too short.

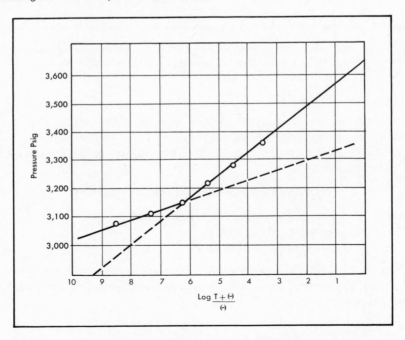

Figure 7–30 Example of an extrapolated shut-in period where the two radii result in the development of two possible lines, indicating a barrier.

TABLE 7–2 DST–PREDICTED GAS-WELL PERFORMANCE COMPARED WITH POST-COMPLETION PERFORMANCE: ASSUMED $n = 0.8$*

Well	Formation	Q_g (MMCFD)	Dr	Predicted Q_1(MMCFD)	Stimulation Treatment	Final Q_1(MMCFD)	Stimulation Effect
White Rose Pembonia 11–26	Blairmore(ss)	2.50	1.42	3.685	1,000 gal. acid	4.50	1.22
H. B. Uno-tex Wimborne 7–29	D-3 (1s)	0.108	35.5	4.00	500 gal. acid	17.20	4.30
Pacific jedney d-77-J	Baldonnel(dol)	1.41	1.7	2.68	4,250 gal. acid	3.50	1.31
Pacific jedney d-77-J	Halfway(ss)	1.25	6.3	9.07	40,000 lb. sand frac.	18.40	2.03
McCoy dome Bubbles b-a62-B	Baldonnel(dol)	0.486	2.1	1.04	11,500 gal. acid	4.80	4.61
Panalta Drumheller 6–3	Glauconitic(ss)	2.50	4.1	11.50	10,000 lb. sand 350 lb. Al. pellet frac.	39.00	3.39

* n = exponent in back-pressure equation, dimensionless.

$$\text{Recoverable stock-tank oil} = \frac{(A)(t)(7{,}758) \times \phi(1 - S_w)R}{FVF}$$

where: A = Area of prospective reservoir, acres
         t = producting formation thickness, ft
    7,758 = bbl/acre-ft
         ϕ = effective porosity, fraction
        S_w = interstitial water, fraction
     FVF = formation volume factor — the ratio of oil removed from reservoir to oil recovered in stock tank after gas has dissipated.
        R = estimated recovery percent expressed as a decimal (see Tables 7-3 and 7-4)

Natural-gas liquid encompasses "those hydrocarbon liquids which are gaseous or in solution with crude oil in the reservoir and which are recoverable as liquids by the process of condensation or absorption which takes place in field separators, scrubbers, etc.[93] The most common type is the wet gas, or gas condensate. Discovery of the probable reserves involves a laboratory exercise and application of the dry-gas calculations. The intricacy involved in solving the reserves problem lies in the fact that natural-gas liquid reserves must be calculated from the parent oil reserve.

$$\text{Recoverable solution gas} = \text{recoverable stock tank oil} \times G_s R_g$$

where: G_s = gas volume liberated from each barrel of in-place oil as it is produced as a barrel of stock-tank oil.
        R_g = recovery factor in decimal form.

In the field where sophisticated laboratory instruments are not available the separator is used to measure the amount of gas produced for each barrel of oil (gas:oil ratio). R_g is estimated and it can be more than the straightforward R as shown in Tables 7-3 and 7-4. Other times it is the same as R and only seldom is it less than R.

Natural-gas reserve estimation is divided into nonassociated gas and associated gas. Nonassociated gas is that gas that is not in contact with crude oil in the particular reservoir. Associated gas is in contact with oil and production of the gas is significantly affected because of the produc-

TABLE 7–3 CHARACTERISTICS OF A RANDOM SELECTION OF RESERVOIRS AND THEIR ESTIMATED OIL RECOVERIES

Producing formation	Average reservoir conditions					Recovery of oil from reservoir					
						Solution gas drive		Gas-cap drive		Water drive	
	Formation condition*	Porosity ϕ	Interstitial water S_w	Formation volume factor FVF	Stock-tank oil in place, bbl/acre-ft F	Bbl/acre-ft F	Recovery factor R	Bbl/acre-ft F	Recovery factor R	Bbl/acre-ft F	Recovery factor R
Miocene............	Good	0.32	0.22	1.35	1,435	460	0.32	574	0.40	900	0.63
	Fair	0.27	0.26	1.35	1,150	320	0.27	356	0.31	600	0.57
	Poor	0.23	0.35	1.35	860	180	0.21	215	0.25	350	0.41
Frio............	Good	0.31	0.24	1.40	1,300	418	0.32	470	0.36	785	0.60
	Fair	0.26	0.30	1.40	1,010	252	0.25	308	0.28	590	0.52
	Poor	0.22	0.40	1.40	732	132	0.18	154	0.21	340	0.46
Hackberry and Lower Frio........ (Nodosaria)	Good	0.28	0.28	1.70	920	296	0.32	331	0.36	535	0.58
	Fair	0.25	0.35	1.70	742	192	0.26	222	0.30	356	0.48
	Poor	0.20	0.45	1.70	502	112	0.22	125	0.25	200	0.40
Cockfield and Yegua............	Good	0.26	0.30	1.45	975	309	0.32	351	0.36	556	0.57
	Fair	0.22	0.35	1.45	765	230	0.30	260	0.34	382	0.50
	Poor	0.18	0.50	1.45	482	133	0.28	150	0.31	203	0.42
Wilcox (upper Gulf Coast)	Good	0.22	0.35	1.80	617	153	0.25	185	0.30	418	0.66
	Fair	0.17	0.45	1.80	403	80	0.20	97	0.24	169	0.42
	Poor	0.13	0.55	1.80	253	43	0.17	51	0.20	86	0.34
Limestones. dolomites, and reefs	Good	0.20	0.35	1.46	694	118	0.17	187	0.27	382	0.55
	Fair	0.16	0.40	1.45	515	76	0.15	106	0.21	204	0.40
	Poor	0.07	0.45	1.45	207	29	0.14	37	0.18	62	0.30

* *Good formation condition:* Uniform blanket formation, thick, not lenticular or broken by shale or impermeable streaks or having sections of poor porosity. If a sand, not highly cemented. If sand, permeabilities of several hundred millidarcys and greater. If limestone, permeabilities of 100 and greater.

Fair formation condition: Intermediate between good and poor, considering all characteristics.

Poor formation condition: Nonuniform, lenticular, thin, or shaly. Poor effective porosity, vugular, or highly cemented. If sand, permeabilities of 100 millidarcys or less. If limestone, permeabilities of 15 millidarcys or less.

TABLE 7-4 GENERALIZED AVERAGE OIL RECOVERIES

Sample No.[*]	Type of formation	Gravity of stock-tank oil, °API	Poros-ity ϕ	Inter-stitial water S_w	Permea-bility, millidarcys	Forma-tion volume factor FVF	Esti-mated recov-ery, bbl/acre-ft F	Recov-ery factor R
			Solution Gas Drive					
1	Sand	38	0.20	0.15	90	1.25	368	0.35
12	Sand	41	0.153	0.40	6.8	1.28†	138	0.25
15	Sand	40	0.17	0.35	125	1.14	187	0.25
18	Sand	32–34	0.202	0.35	395	1.45	237	0.34
36	Sand	39	0.162	0.265	25–228	1.14	354	0.33
57	Sand	46.4	0.28	0.25	600	1.18	460	0.33
67	Sand	44.2	0.255	0.30	300	1.47	343	0.36
69	Sand	40	0.218	0.25	400–500	1.10	285	0.25
70	Sand	22.5	0.22	0.30	500	1.10	453‡	0.42
86	Sand	33.1	0.23	0.35	500	1.03	225	0.20
94	Sand	36.5	0.235	0.292	130	1.50	126	0.15
95	Sand	36.0	0.222	0.31	85	1.54	180	0.23
			Gas-cap Drive					
5	Sand and lime	41	0.125	0.20	150	1.24	232	0.37
19	Sand	42	0.18	0.32	80	1.10	297	0.34
21	Sand	31	0.20	0.20	400	1.06	530	0.45
22	Sand	37.2	0.182	0.25	356	1.10	213	0.22
24	Sand	41	0.13	0.20	206	1.67	233	0.48
25	Sand	38.5	0.15	0.20	....	1.25	247	0.33
28	Sand	32	0.179	0.35	100	1.08	180	0.22
42	Sand	44	0.27	0.35	....	1.25	547‡	0.50
46	Sand	38.6	0.18	0.38	100	1.15	256	0.34
47	Sand and lime	38	0.18	0.40	90	1.28	165	0.25
48	Sand and lime	37	0.172	0.355	75	1.20	248	0.35
49	Sand	22.5	0.23	0.25	....	1.05†	500‡	0.39
50	Sand	24.5	0.29	0.20	....	1.08†	443‡	0.27

SOURCE: From a tabulation of 103 samples of an American Petroleum Institute study by Craze and Buckley, A Factual Analysis of the Effect of Well Spacing on Oil Recovery, in "Drilling and Production Practices," American Petroleum Institute, 1945.

* Sample numbers refer to the original publication.

† Estimated.

‡ Well spacing of 10 acres or less.

TABLE 7-4 (*Continued*)

Water Drive

2	Sand	31	0.30	0.35	1,000	1.20	600	0.47
3	Sand	31	0.18	0.20	75	1.07	410	0.55
6	Sand	38–39.8	0.252	0.17	2,000–3,000	1.20	1,058	0.78
10	Sand	23	0.32	0.30	50–5,000	1.20	629	0.43
16	Sand	35	0.34	0.33	3,000	1.29	958	0.70
20	Sand	39	0.14	0.25	118	1.07	306	0.40
23	Lime	39	0.176	0.30	1,010	1.49	407	0.73
26	Sand	33.2	0.15	0.15	328	1.1	437	0.49
38	Sand	20	0.28	0.10	5,000	1.09†	1,121	0.61
43	Sand	39	0.269	0.259	1,335	1.51	617	0.60
44	Sand	36.6	0.179	0.40	104	1.04	429	0.53
89	Sand	37	0.327	0.20	1,865	1.43	1,030	0.73

tion of crude oil. At initial reservoir conditions the following formula applies:

$$\text{Recoverable gas (ft}^3) = \left[43{,}560 \times At\phi(1 - S_w) \frac{460 + 60}{460 + T_r} \frac{P_r}{P_b} \frac{1}{Z_r} \right]$$
$$- \left[43{,}560 \times At\phi(1 - S_w) \frac{460 + 60}{460 + T_r} \frac{P_r}{P_b} \frac{1}{Z_a} \right] \qquad (7\text{-}1)$$

The first half of the right side of the equation represents the initial reservoir conditions and the second half represents the abandonment reservoir conditions. Another formula (utilizing the same symbols) employing the factor R (recovery factor) is:

$$\text{Recoverable gas (ft}^3) = 43{,}560 \, (At\phi)(1 - S_w) \frac{460 + 60}{460 + T_r} \frac{P_r}{P_b} \frac{1}{Z_r} R \qquad (7\text{-}2)$$

where:

$43{,}560 = \text{ft}^3/\text{acre-ft}$
$460 = $ amount added to convert from °F. to °R.
$A = $ area of prospective reservoir, acres
$t = $ mean reservoir thickness, ft
$\phi = $ effective porosity, fraction
$S_w = $ interstitial water content, fraction
$T_r = $ average reservoir temperature, °F.
$P_b = $ standard gas-measurement pressure base, 14.73 psia
$P_r = $ average initial reservoir pressure, psia
$Z_r = $ compressibility factor to get original reservoir conditions

Z_a = compressibility factor at abandonment reservoir conditions
P_a = average reservoir pressure at abandonment reservoir conditions
R = recovery factor, fraction

Equation 7-1 applies to reservoirs which have no water drive and in which the reservoir pore space occupied by gas will remain constant. Equation 7-2 applies to formations which are actively influenced by a strong water drive. The water drive, through the course of the production period, alters the pore space and therefore the recoverable gas cannot be measured by relating the original pressure to an assumed abandonment pressure.

The abandonment pressure (without water drive) depends on the commercial viability of production. Between 700 and 1,000 psi will be used by operators except where compressors are installed and a figure of 300–500 psi is a reasonable pressure bracket from which to determine abandonment. Recovery of gas is very high when compared to the recovery of oil. Gas recovery can be as high as 70–85%.

The following problem serves to illustrate how the amount of recoverable gas is calculated. Equation 7-1 is used where:

Recoverable gas (ft³) = Initial Reservoir Condition
                        − Reservoir Abandonment Condition

GIVEN:
Reservoir area	A = 500 acres
Reservoir thickness	t = 40 ft
Porosity	ϕ = 30% or 0.30
Interstitial water	S_w = 40% or 0.40
Reservoir temperature	T_r = 160°F.
Measurement pressure base	P_b = 14.65 psia
Initial reservoir pressure	P_r = 4,000 psia
Abandonment pressure	P_a = 700 psia
Specific gravity of gas	= 0.72

Step 1.

$$RG \text{ (ft}^3) = \left[(43,560)(500)(40)(0.30)(1 - 0.40) \left(\frac{460 + 60}{460 + 160} \right) \left(\frac{4,000}{14.65} \right) \left(\frac{1}{Z_r} \right) \right]$$

$$- \left[(43,560)(500)(40)(0.30)(1 - 0.40) \left(\frac{460 + 60}{460 + 160} \right) \left(\frac{700}{14.65} \right) \left(\frac{1}{Z_a} \right) \right]$$

To find Z_r and Z_a, enter Fig. 7-31 at the given specific gravity for gas. Find the pseudocritical temperature by entering the bottom half of the

Compressibility factors for natural gas

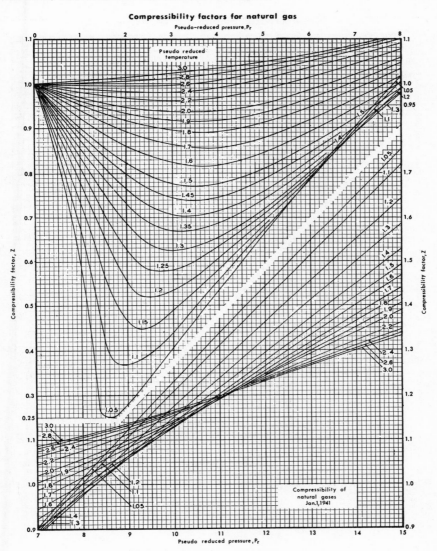

Figure 7–31 Compressibility factors for natural gas

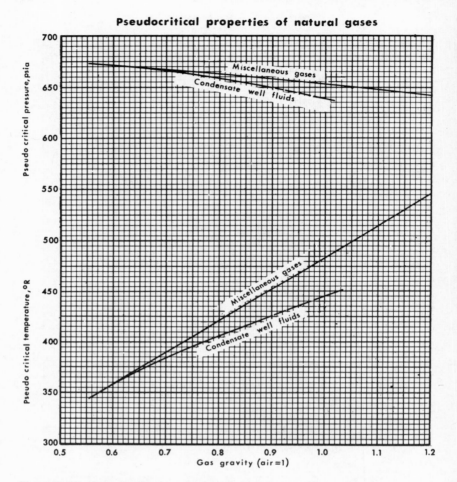

Figure 7–32 Pseudocritical properties of natural gas

chart and the pseudocritical pressure by entering the top half of the chart. They are 386°R. and 665 psia respectively. Apply the pseudocritical figures to the following formulas.

$$\text{Pseudoreduced pressure} = \frac{\text{Initial reservoir pressure } (P_r)}{\text{pseudocritical pressure}}$$

$$\text{Pseudoreduced temperature} = \frac{460 + \text{reservoir temp } (T_r)}{\text{pseudocritical temp.}}$$

$$\text{Pseudoreduced pressure} = \frac{4,000}{665} = 6.0$$

$$\text{Pseudoreduced temperature} = \frac{620}{386} = 1.6$$

Enter Fig. 7–32 at the top at pseudoreduced pressure = 6 and intersect with the pseudoreduced temperature contour 1.6. This yields a compressibility factor of 0.89 which applies to initial reservoir conditions. Thus Z_r is solved. Z_a is solved in the same fashion and the original problem can be completed.

Diamond Core Drilling

Successful diamond drilling requires that the energy level of the mud flow through the bit is high to ensure adequate cutting removal beneath the bit and efficient cooling of the diamonds. The level of hydraulic energy at the bit is measured in horsepower per square inch (hsi).

In hard formations, where the cooling and lubrication of the diamonds is more important than the removal of cuttings, 2.0–2.5 hsi is generally sufficient to achieve optimum performance.

In softer formations, where a larger volume of cuttings has to be removed, 2.5 to 3.0 hsi is recommended.

The energy level of hsi can be calculated from Fig. A–1, knowing the mud volume through the bit in gpm, the bit pressure drop in psi and the hole size in inches.

The bit pressure drop can either be evaluated by the difference of pressure when the bit is on and off bottom, or read from Fig. A–2, the "Bit Pressure Drop Chart," using the mud channels factor (TFA) stamped on all high-pressure-drop Christensen diamond bits.

"Pump off" is a result of the hydraulic forces between the bit face and the formation which tends to lift the bit off bottom. This force becomes appreciable in high-pressure-drop bits. Fig. A–3 must be consulted to determine the additional bit weight required to compensate for the hydraulic "pump off" force.[108]

Recommended rotating speeds, bit weights, and pump-discharge rates for a variety of drilling conditions are given in Fig. A–4.

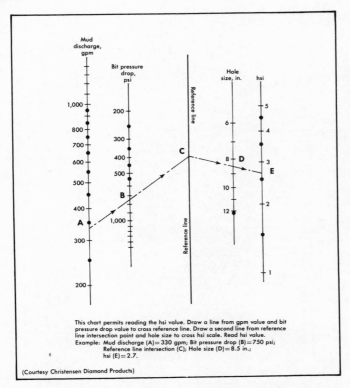

This chart permits reading the hsi value. Draw a line from gpm value and bit pressure drop value to cross reference line. Draw a second line from reference line intersection point and hole size to cross hsi scale. Read hsi value.

Example: Mud discharge (A)=330 gpm; Bit pressure drop (B)=750 psi; Reference line intersection (C); Hole size (D)=8.5 in.; hsi (E)=2.7.

(Courtesy Christensen Diamond Products)

Figure A–1 Nomograph for finding hydraulic horsepower per square inch (hsi).

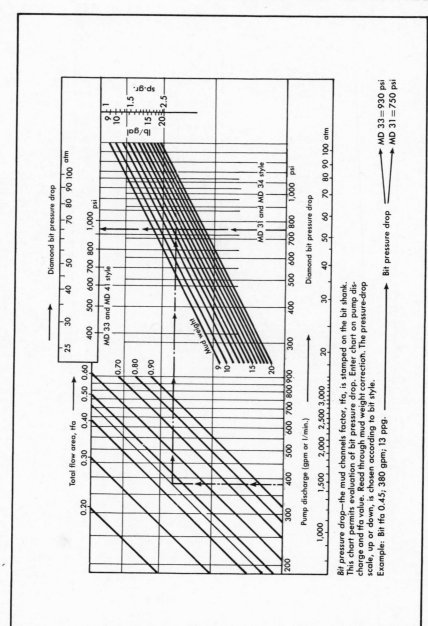

Bit pressure drop—the mud channels factor, tfa, is stamped on the bit shank. This chart permits evaluation of bit pressure drop. Enter chart on pump discharge and tfa value. Read through mud weight correction. The pressure-drop scale, up or down, is chosen according to bit style.
Example: Bit tfa 0.45; 380 gpm; 13 ppg.

Figure A-2   Diamond bit pressure-drop chart.

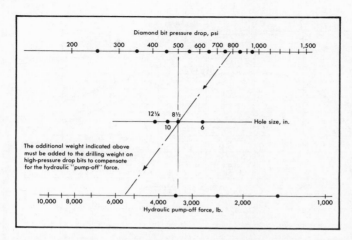

Figure A-3 Nomogram to determine additional pump pressure required to compensate for the "pump off" force. Courtesy of Christensen Diamond Products.

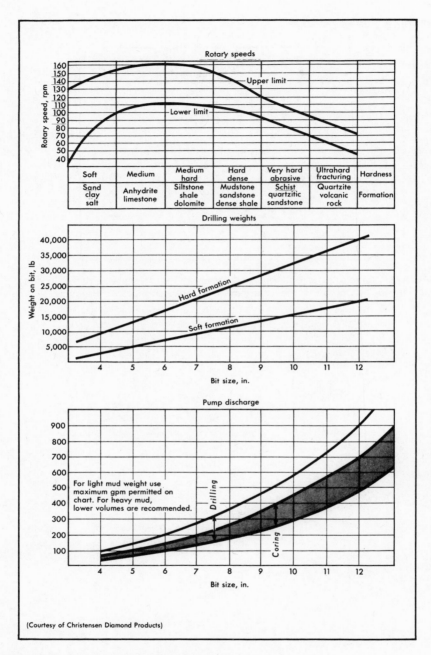

Figure A–4 Some recommended drilling practices.

B

FIGURE B–1 DENSITY OF MERCURY AT VARIOUS TEMPERATURES

Temp., °C.	Mass, g/cc	Temp., °C.	Mass, g/cc
0	13.5955	26	13.5315
1	593	27	529
2	590	28	526
3	588	29	524
4	585	30	13.521
5	13.583	31	519
6	580	32	516
7	578	33	514
8	575	34	511
9	573	35	13.509
10	13.570	36	507
11	568	37	504
12	565	38	502
13	563	39	499
14	561	40	13.497
15	13.558		
16	556		
17	553		
18	551		
19	548		
20	13.546		
21	543		
22	541		
23	538		
24	536		
25	13.534		

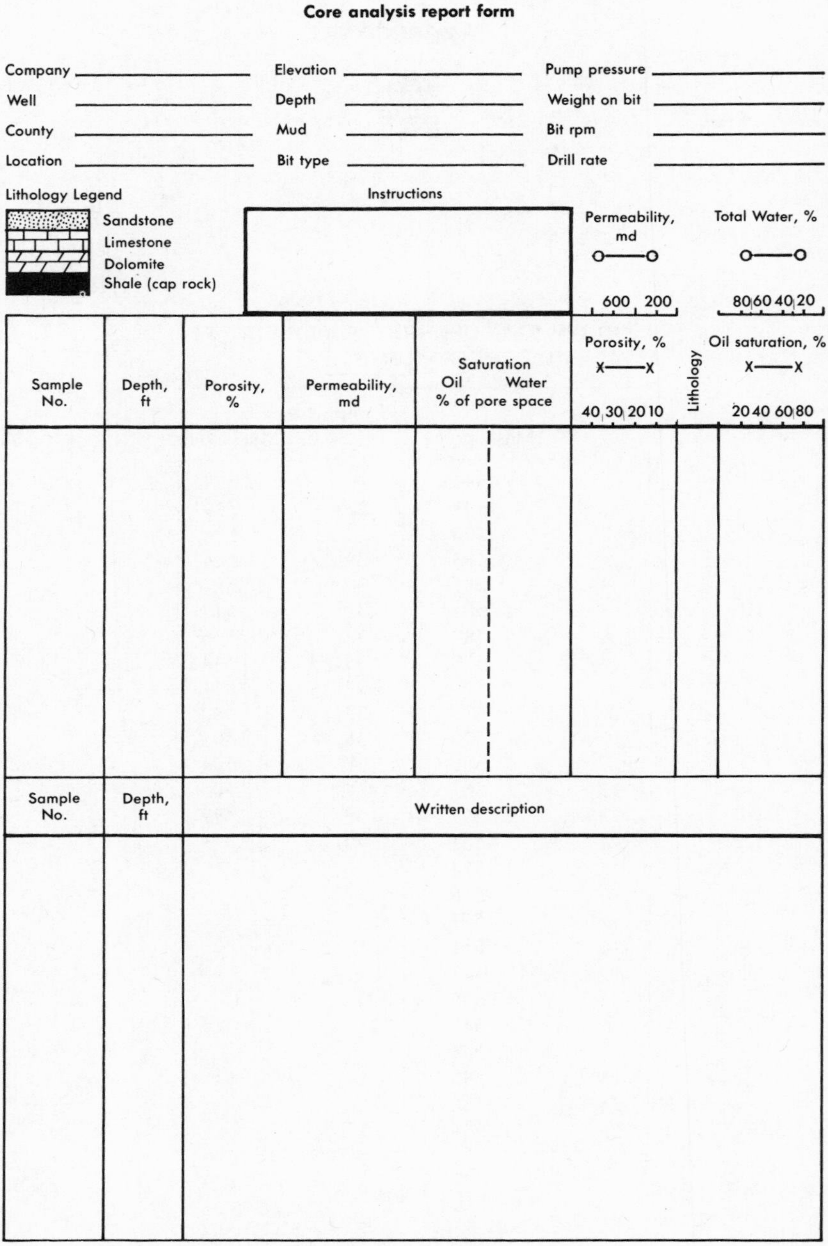

Core analysis report form

Company _____	Elevation _____	Pump pressure _____
Well _____	Depth _____	Weight on bit _____
County _____	Mud _____	Bit rpm _____
Location _____	Bit type _____	Drill rate _____

Lithology Legend

Sandstone
Limestone
Dolomite
Shale (cap rock)

Instructions

Permeability, md
o———o
600   200

Total Water, %
o———o
80 60 40 20

Sample No.	Depth, ft	Porosity, %	Permeability, md	Saturation Oil Water % of pore space	Porosity, % X———X 40 30 20 10	Lithology	Oil saturation, % X———X 20 40 60 80

Sample No.	Depth, ft	Written description

Figure B–1 Typical core-analysis report form.

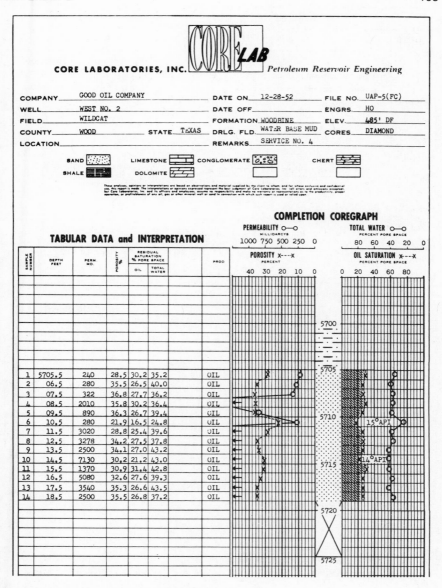

Figure B-2 Typical Core Laboratories report form.

1	2	3	4	5	6
	Measured				
Depth	Wgt. Saturation Sample w/Fluids, g	Recovered Free H$_2$O cc	Recovered, Oil cc °F API	Oil Gravity w/temp corr. from table, API	Corrected Oil Volume from graph, cc
1 2					

Formula No. 1

Pore Volume $\dfrac{\text{(Bulk Vol of Porosity Sample)(Wgt. Sat. Sample w/Fluids)(Porosity)}}{\text{Wgt. Porosity Sample w/Fluids}}$

Formula No. 2

%H$_2$O Saturation $\dfrac{\text{(Recovered Free H}_2\text{O)(100)}}{\text{Pore Volume of Saturation Sample}}$

Formula No. 3

% Oil Saturation $\dfrac{\text{(Corrected Oil Volume)(100)}}{\text{Pore Volume of Saturation Sample}}$

AIR PERMEABILITY

1	10	11	12	13	14	15
	Measured					
Depth	Length (cm) L	Diam (cm) D	Flowmeter Reading (cm)	Flowmeter (Small) (Medium) (Large)	Air Temperature (°C)	Outflow Rate from graph cc/sec Q
1 2						

MERCURY-INJECTION

1	20	21	22	23	24	25
	Measured					
Depth	Wgt. Porosity Sample w/Fluids, g	Wgt. Porosity Sample Dry, g	Bulk Volume cc	Pore Volume, cc	Grain Volume cc	Grain Density, sp. gr.
1 2						

Specific Gravity

Ss   2.65 average   Grain Volume = Bulk Volume − Pore Volume
Ls   2.72   "
Dol 2.86   "

PROCEDURE

	Calculated		
7	8	9	
Pore Volume, from Formula No. 1, cc	Water Saturation from Formula No. 2, %	Oil Saturation from Formula No. 3, %	Remarks

Supplementary Formulae

$$\text{Bulk Vol. Sat. Sample} = \frac{(\text{Wgt of Sat. Sample})(\text{Bulk Vol. of Porosity Sample})}{\text{Wgt Porosity Sample w/Fluids}}$$

$$\text{Pore Volume} = \frac{(\% \text{ Porosity})(\text{Bulk Vol. of Sat. Sample})}{100}$$

PROCEDURE

	Calculated		
16	17	18	19
Air Viscosity from graph, cps μ	Cross-section Area from graph, sq cm A	Darcys Small $\quad K = \dfrac{\mu QL}{A}$ Medium $\quad K = \dfrac{2QL}{A}$ Large $\quad K = \dfrac{4QL}{A}$	Millidarcys

POROSITY PROCEDURE

	Calculated
26	27
Effective Porosity, %	Sample Description

$$\text{Grain Density} = \frac{\text{Dry Wgt of Porosity Sample}}{\text{Grain Volume}} \text{ (in g/cc)}$$
(Bulk Density)

$$\text{Effective Porosity} = 100 \times \frac{\text{Pore Volume}}{\text{Bulk Volume}}$$

	Measured				
1	28	29	30	31	32
Depth	Reference Volume, cc (Vf)	Reference Volume, cc (V_g)	Wgt. Porosity Sample w/Fluids, g	Wgt. Porosity Sample Dry, g	Sample Reference Volume cc (V_f)
1 2					

	Measured				
1	39	40	41	42	43
Depth	Wgt. Porosity Sample w/Fluids, g	Wgt. Porosity Sample Dry, g	Bulk Volume, cc	Volumeter w/o Sample, cc	Volumeter w/Sample, cc
1 2					

Figure B–3 Measurement and calculation work sheet.

POROSITY PROCEDURE

	Calculated				
33	34	35	36	37	38
Bulk Volume	Grain Volume from graph, cc (V_g)	Pore Volume, cc	Grain Density, sp. gr.	Effective Porosity, %	Sample Description

Pore Volume = Bulk Volume − Grain Volume

POROSITY PROCEDURE

	Calculated			
44	45	46	47	48
Grain Volume, cc 43 − 42	Pore Volume, cc	Grain Density, sp. gr.	Total Porosity, %	Sample Description

Time, min	Vol. rec., H_2O		Time, min	Vol. rec., H_2O		Time, min	Vol. rec., H_2O
1.0			1.0			1.0	
2.0			2.0			2.0	
3.0			3.0			3.0	
15.0			15.0			15.0	
Final			Final			Final	
Oil			Oil			Oil	
1.0			1.0			1.0	
2.0			2.0			2.0	
3.0			3.0			3.0	
14.0			14.0			14.0	
15.0			15.0			15.0	
Final			Final			Final	
Oil			Oil			Oil	

Figure B–4 Saturation data sheet.

Figure B–5 Temperature corrections to readings of API hydrometers for American Petroleum Oils at various temperatures (Standard at 60°F.; modulus 141.5).

Observed temperature °F.	Observed degrees API.					
	20.0	30.0	40.0	50.0	60.0	70.0
	Add to observed degrees API.					
30	1.7	2.0	2.4	3.0	3.7	4.3
32	1.6	1.9	2.3	2.8	3.4	4.0
34	1.5	1.8	2.1	2.6	3.1	3.7
36	1.4	1.6	2.0	2.4	2.9	3.4
38	1.3	1.5	1.8	2.2	2.6	3.1
40	1.2	1.4	1.6	2.0	2.4	2.8
42	1.1	1.2	1.5	1.8	2.2	2.5
44	.9	1.1	1.3	1.6	2.0	2.2
46	.8	.9	1.1	1.4	1.7	1.9
48	.7	.8	.9	1.2	1.4	1.6
50	.6	.7	.8	1.0	1.2	1.4
52	.5	.6	.7	.8	1.0	1.1
54	.3	.4	.5	.6	.8	.9
56	.2	.3	.3	.4	.5	.6
58	.1	.1	.1	.2	.3	.3
	Subtract from observed degrees API.					
60	.0	.0	.0	.0	.0	.0
62	.1	.1	.1	.2	.2	.3
64	.2	.3	.3	.4	.4	.6
66	.3	.4	.5	.6	.7	.8
68	.5	.6	.6	.7	.9	1.1
70	.6	.7	.8	.9	1.1	1.4
72	.7	.8	.9	1.1	1.3	1.6
74	.8	.9	1.1	1.3	1.6	1.8
76	.9	1.1	1.3	1.5	1.8	2.1
78	1.0	1.2	1.4	1.7	2.0	2.4
80	1.1	1.3	1.5	1.8	2.2	2.6
82	1.2	1.4	1.7	2.0	2.5	2.9
84	1.3	1.5	1.8	2.2	2.7	3.2
86	1.4	1.7	2.0	2.4	2.9	3.4
88	1.6	1.8	2.1	2.6	3.1	3.7
90	1.7	2.0	2.3	2.7	3.3	3.9
92	1.8	2.1	2.4	2.9	3.5	4.2
94	1.9	2.2	2.6	3.1	3.8	4.4
96	2.0	2.3	2.7	3.3	4.0	4.6
98	2.1	2.4	2.9	3.4	4.2	4.9
100	2.2	2.6	3.0	3.6	4.4	5.1

$$\text{API. at } 60°F. = \frac{141.5}{\text{Specific gravity at } 60°F. \text{ relative to water at } 60°F.} - 131.5$$

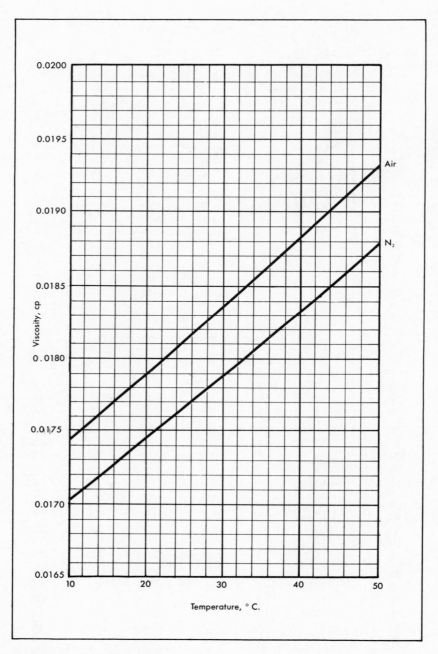

Figure B–6 Viscosity curves.

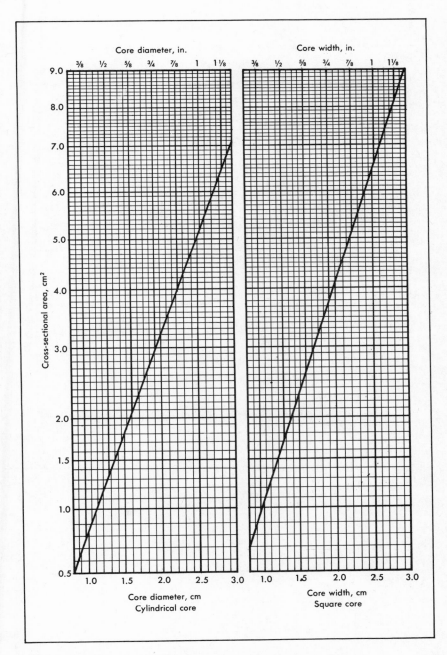

Figure B-7 Diameter-area curves for permeability plugs.

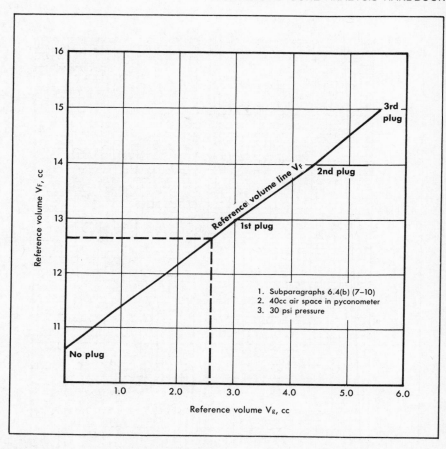

Figure B-8 Air porosity measurement graph.

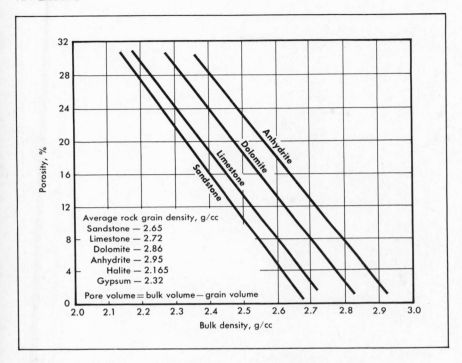

Figure B-9 Porosity vs. bulk density for various sedimentary rocks.

Appendix

C

History divisions				Time, millions of years
Phanerozoic eon	Cenozoic era	Pleistocene epoch		1
		Pliocene epoch	Tertiary period	13
		Miocene epoch		25
		Oligocene epoch		36
		Eocene epoch		58
		Paleocene epoch		65
	Mesozoic era	Cretaceous period		135
		Jurassic period		180
		Triassic period		230
	Paleozoic era	Permian period		280
		Pennsylvanian period		
		Mississippian period		350
		Devonian period		405
		Silurian period		425
		Ordovician period		500
		Cambrian period		600

Figure C–1 Geologic time table

FIGURE C-2 TECTONO-ENVIRONMENTAL CLASSIFICATION OF SEDIMENTARY ROCKS[107]

	Fluvial-lacustrine-eolian environment	Transitional environment (fluvial-lagoonal-littoral)	Epineritic environment	Infraneritic environment
General conditions; types of specific environments	Alluvial plains, stream channels, lakes, swamps, local or extensive wind action.	Alluvial plains, lagoons, marshes or swamps, barrier beaches, deltaic conditions.	Shallow wave and current-agitated marine waters, open circulation.	Water depths exceed 120 feet; relatively quiet off shore zones.
General colors and properties of sediments; faunal elements	White, gray, yellow red, brown, maroon, mottled, Lenticular or sheet sandstones. Blocky to poorly bedded shales. Limestones very subordinate; coal and lignite beds. Plant impressions and remains; land vertebrates; freshwater gastropods and pelecypods.	Gray brown red, green, blue, dark gray, and black. Lenticular sheet stones; some sheet sands. Shales blocky to well bedded; coal and lignite beds. Limestones subordinate; here and there a tongue of marine limestone. Plants land vertebrates, gastropods, pelecypods, phosphatic brachiopods, and ostracodes common.	Gray, light brown greenish gray, bluish gray, dark gray to black. Sheet sandstones; local linear bodies. Shales commonly bedded; limestones range from argillaceous to fossiliferous-fragmental types. Great variety of stout-shelled benthonic invertebrates (mollusks, brachiopods, echinoderms, corals, etc.).	Sediment colours gray, greenish gray, brownish gray, dark gray, black. Reds and browns subordinate. Sheet sands mainly finegrained; shales well bedded. Limestones show wide variety; normal marine types dominant. Great variety of benthonic and nektobenthonic types, including more delicate forms. Significant per cent planktonic types.
Stable shelf occurrence	Quartzose sandstones, crossbedded. Massive clay shales, commonly mottled. Carbonaceous; seldom calcareous. Freshwater limestones, nodular, dense, local.	Quartzose sandstones, crossbedded, commonly lenticular. Clay shales dominant, brackish varieties bedded; carbonaceous, locally calcareous, marly. Freshwater limestone subordinate.	Quartzose sandstone cross-bedded. Siltstones ripplemarked. Shales commonly fine clayey, and greenish. Lms. with clastic textures, evenly bedded or locally crossbedded. Ss. may grade directly to lms.	Fine-grained quartzose sandstones. Clay shales common, well bedded; calcareous, carbonaceous. Limestones normal marine, fusulinid, chalk; numerous planktonic components.
Unstable currence	Lenticular quartzose sandstones; sheet sands are subgraywacke. Shales mainly siltstone; claystones commonly bedded; micaceous, carbonaceous, calcareous. Fresh-water to marly limestones.	Subgraywacke sandstones, some linear quartzose channel sands. Shales mainly siltstones, massive to banded, mic., carb., seldom calc. Fresh-water lms. subordinate, nodular dense to uneven sugary texture.	Quartzose to subgraywacke sandstones, cross-bedded, ripplemarked. Shales commonly siltstones, carbonaceous, calcareous, light colors. Lms. thicker stable neritic types, argillaceous, locally denser.	Fine-grained to subgraywacke sandstones, evenly bedded. Silty claystones carbonaceous, calcarious. Limestones as above, locally denser, less widespread.

Epineritic biostromal environment	Bathyal-abyssal environment	Restricted lagoonal humid	Restricted lagoonal arid
Shallow clear waters. open circulation. little or no land-derived sediment. Controlling factors: temp.. salinity, oxygen content. depth.	Water depths exceed 600 feet; these conditions may locally be fulfilled in rapidly subsiding geosynclines. Rare or absent in craton.	Mainly neritic depths; volume may be constant, but circulation restricted by barriers, sills, or biohermal control.	Mainly epineritic depths; evaporation exceeds inflow. Circulation restricted by barriers, sills, or biohermal control.
Light colors, with tan bluish. cream dominant. Sandstones and shales subordinate; main bulk of sediments are carbonates. with abundant evidence of life forms and associated debris. Corals. bryozoans. algae. oysters. specialized brachiopods and larger foraminifera, crinoids. etc.	Sediment colors commonly dark; blue. green, red. dark gray to black. Land-derived sediments relatively rare; shales siliceous. Chiefly planktonic types. smaller foraminifera, diatoms, pteropods, etc.	Sediment colors commonly dark gray to black. Sandstones rare; shales dominant; bituminous. thin-bedded. Phosphatic brachiopods, conodonts, certain mollusks, spores. algae.	Sediment colors commonly light; white. cream. brownish. greenish. bluish. pink. red. Sandstones rare. Clay shales dominate. gypsiferous. calcareous. Limestones dense; primary dolomites. nodular to thin-bedded. Evaporites may range from subordinate to dominant. Fauna aberrant. depauperate or lacking.
This is a typical environment of stable to mildly unstable shelves, in widespread shallow seas. The biostromal area may occur locally or over large areas. Sporadic bioherms common. Limestones includes fossiliferous-fragmental. reefoid, oölitic. and chalks. Subordinate clay shales marls. and thin quartzose sands in the association.	These environmental conditions doubtfully present on shelf areas; may occur locally in intracratonic basins. Typical but rare occurrence in eugeosynclinal association at times of rapid subsidence and slow deposition. No sandstones known; shales very siliceous splintery; primary chert beds or nodules; limestone dark. dense. very siliceous.	In shelf occurrences. this environment may form widespread black shales behind barrier beaches or fringing reefs. The sequence may grade into continental sands and shales on one or more sides. Within the environment sandstones and limestones are rare.	In shelf occurrences this environment may form widespread evaporitic sequences dominated by gypsiferous shale and thin evaporite beds. grading. landward to typical redbed continental and transitional sequences. Limestones thin and subordinate within sequence.
Under basin or geosynclinal conditions. biostromes may rarely occur in local optimum areas. Significant is the growth of biohermal zones along tectonic hinge lines at edges of intracratonic basins. Here reefoid. frag-		Thick sequences of dark shales. bituminous. waxy. mainly clay types. Sandstones rare. fine-grained. Limestones subordinate. dark. bituminous varieties.	Thick sequences of typical evaporite associations; gypsum. anhydrite. salt. thin limestones. and dolomites; bright-colored shales. commonly clayey and gypsiferous. Cyclical evaporites common.

	Fluvial-lacustrine-eolian environment	Transitional environment (fluvial-lagoonal-littoral)	Epineritic environment	Infraneritic environment
Intracratonic basin occurrence.	Subgraywacke sandstones, local arkose associations. Shales mainly uneven-textured siltstones; micaceous, carbonaceous, semiwaxy, seldom calcareous. Lms. nodular to uneven thin-bedded.	Quartzose, subgraywacke, arkosic sandstones; thick shales commonly change characteristics from top to bottom. Micaceous, carbonaceous, calcareous. Freshwater lms. may locally be thick.	Subgraywacke sandstones, arkosic sheets or wedges. Shales silty, carbonaceous, micaceous, calcareous. Limestones, nodular, uneven, may be argillaceous or dense.	Subgraywacke sandstones, thinbedded. Shales fine siltstones, to clay shales dark colors common. Limestones thin, commonly dark; may be nodular.
Geosynclinal occurrence	Graywacke sandstones with subordinate subgraywacke channels. Shales massive to banded, chloritic or feldspathic. Freshwater lms. may form thick lenses.	Graywacke sandstones with subordinate subgraywackes. Massive to banded shales, mainly silty to sandy. Local thick freshwater limestones.	Graywacke sandstones with thinner subgraywackes. Shales commonly uneven-bedded; may show dark colors. Limestones subordinate, nodular.	Graywacke sandstones subordinate. Shales mainly siltstones with uneven texture. Limestones rare; siliceous, dark.

Epineritic biostromal environment	Bathyal-abyssal environment	Restricted lagoonal humid	Restricted lagoonal arid
mental, oolitic lime-stones are the rule, with other types sub-ordinate.		Dark gray, brown greenish silty shales, uneven textures. Some fine graywackes or sub-graywackes sand-stones. Limestones rare or absent.	It is doubtful whether true evaporates basins occur in geosynclinal conditions, although local sheets of evap-oritic beds and gypsiferous shales may be associated with deltaic parts of geo-synclinal sedimenta-tion.

Simplified for ease of reference. Adapted from W. C. Krumbein, L. L. Sloss, and E. C. Dapples, Sedimentary Tectonics and Environments, Bull, AAPG, vol. 33, no. 11, pp. 1876–1877, 1949. Abbreviations: mic., micaceous; calc., calcareous; carb., carbonaceous; lms., limestone; ss., sandstone. (From G. B. Moody, Petroleum Exploration Handbook, Appendix A, Table A-1, Copyright © 1961. Used with permission of McGraw-Hill Book Co.)

Figure C-2 Tectono-environmental classification of sedimentary rocks.[97]

FIGURE C–3   GRAIN DENSITIES

Anhydrite	2.95 g/cc
Dolomite	2.85
Calcite	2.71
Limestone	2.70
Quartz	2.66
Sandstone	2.65
Kaolinite	2.63
Illite	2.76
Montmorillonite	2.00
Halite	2.17
Coal	1.00–1.80

Appendix

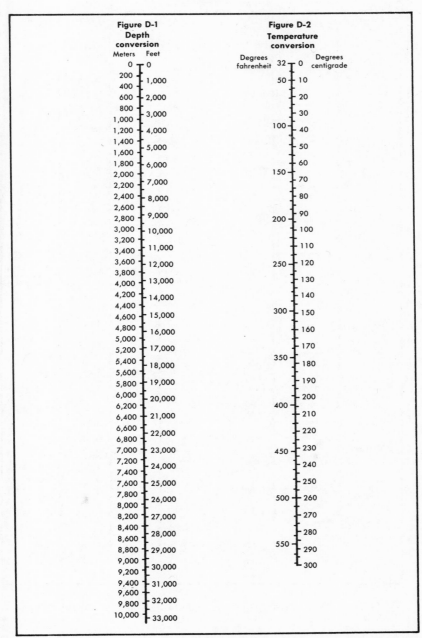

Figure D–1 Depth conversion.   **Figure D–2** Temperature conversion.

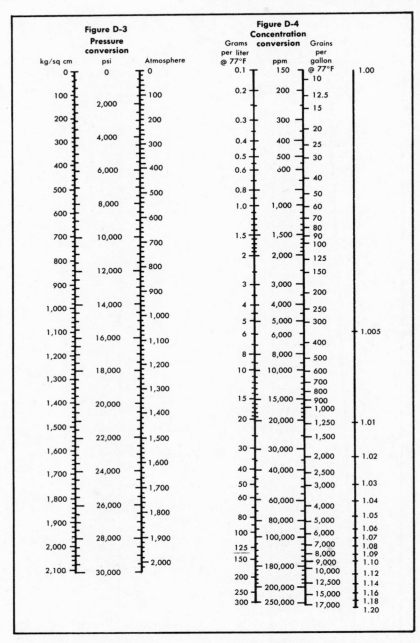

Figure D–3   Pressure conversion.   **Figure D–4**   Concentration conversion.

FIGURE D-5 OIL-GRAVITY CONVERSION

°API	Sp. gr.	Lb/cu ft
8	1.014	63.4
10	1.000	62.5
12	0.986	61.6
14	0.973	60.8
16	0.959	60.0
18	0.946	59.1
20	0.934	58.4
22	0.922	57.6
24	0.910	56.9
26	0.898	56.1
28	0.887	55.4
30	0.876	54.7
32	0.865	54.1
34	0.855	53.4
36	0.845	52.8
38	0.835	52.2
40	0.825	51.6
42	0.815	51.0
44	0.806	50.4
46	0.797	49.8
48	0.788	49.3
50	0.780	48.7
52	0.771	48.2
54	0.763	47.7
56	0.755	47.2
58	0.747	46.7
60	0.739	46.2

$$°API @ 60°F = \frac{141.5}{G} - 131.5$$

G = Specific Gravity @ 60/60° F.

FIGURE D–6 CONVERSION TABLE FOR MUD
GRADIENT

Gradient, psi/1,000 ft of depth	lb/gal	Density lb/cu ft	Sp. gr.
433	8.3	62.4	1.00
440	8.5	63.4	1.02
450	8.7	64.8	1.04
460	8.9	66.2	1.06
470	9.1	67.7	1.09
480	9.2	69.1	1.11
490	9.4	70.6	1.13
500	9.6	72.0	1.15
510	9.8	73.4	1.18
520	10.0	74.9	1.20
530	10.2	76.3	1.22
540	10.4	77.8	1.25
550	10.6	79.2	1.27
560	10.8	80.6	1.29
570	11.0	82.1	1.32
580	11.2	83.5	1.34
590	11.4	85.0	1.36
600	11.6	86.4	1.39
610	11.7	87.8	1.41
620	11.9	89.3	1.43
630	12.1	90.7	1.45
640	12.3	92.2	1.48
650	12.5	93.6	1.50
660	12.7	95.0	1.52
670	12.9	96.5	1.55
680	13.1	97.9	1.57
690	13.3	99.4	1.59
700	13.5	100.8	1.62
710	13.7	102.2	1.64
720	13.9	103.7	1.66
730	14.1	105.1	1.69
740	14.3	106.6	1.71
750	14.4	108.0	1.73
760	14.6	109.4	1.76
770	14.8	110.9	1.78
780	15.0	112.3	1.80
790	15.2	113.8	1.82
800	15.4	115.2	1.85
810	15.6	116.6	1.87
820	15.8	118.1	1.89
830	16.0	119.5	1.92
840	16.2	121.0	1.94
850	16.4	122.4	1.96

FIGURE D–6 (*Continued*)

Gradient, psi/1,000 ft of depth	lb/gal	Density lb/cu ft	Sp. gr.
860	16.6	123.8	1.99
870	16.8	125.3	2.01
880	16.9	126.7	2.03
890	17.1	128.2	2.06
900	17.3	129.6	2.08
910	17.5	131.0	2.10
920	17.7	132.5	2.12
930	17.9	133.9	2.15
940	18.1	135.4	2.17
950	18.3	136.8	2.19
960	18.5	138.2	2.22
970	18.7	139.7	2.24
980	18.9	141.1	2.26
990	19.1	142.6	2.29

FIGURE D-7   USEFUL CONVERSION CONSTANTS

Multiply	By	and Obtain
Barrels	5.6146	Cu ft
Barrels	9,702.03	Cu in.
Barrels	42.0	Gal. (US)
Cu Ft	1,728	Cu in.
Cu Ft	.037037	Cu yard
Cu Ft	7.48055	Gal
Cu Ft	.1781	Bbl (42 gal)
Cu ft/min	.1781	Bbl/min
Cu ft/min	10.686	Bbl/hr
Cu ft water	62.422	Pounds
Cu in.	.0005787	Cu ft
Cu in.	.004329	Gal
Cu in.	.0001031	Bbl
Cu yard	27.0	Cu ft
Cu yard	201.974	Gal
Gal water	8.33	Pounds
Gal	.13368	Cu ft
Specific gravity	62.4	Lb/cu ft
Specific gravity	8.33	Lb/gal
Feet	.30481	Meters
Feet	3.0481	Decimeters
Feet	30.481	Centimeters
Inch	2.54	Centimeters
Inch	25.4	Millimeters
Miles	5,280	Feet
Miles/hr	1.4666	Feet/sec.
Sq in.	.006944	Sq ft
Sq ft	144	Sq in
Pounds water	.11984	Gal. water
Pounds water	.01602	Cu ft water
Horsepower	33,000	Ft lb/min
Horsepower	550	Ft lb/sec

1 Atmosphere = 14.7 psi
°Farenheit = (1.8)(°Centigrade + 32)
°Centigrade = 5/9(°Farenheit −32)
1 acre-foot = 7,758 bbl = 43,560 cu ft

FIGURE D-8 USEFUL FORMULAS

Formula for Stuck Pipe

$$SPL = 735 \times (10)^3 \frac{(W)\,(L)}{(F_2 - F_1)}$$

where:

SPL = stuck pipe location
W = pipe weight (nominal . . . in drilling mud)
L = length of stretch (inches)
F_2 = force (lb) when pipe is stretched.
F_1 = force (lb) of pipe in tension.

Annular Velocity Formula

$$\text{Annular velocity} = \frac{\text{Pump output (bbl/min)}}{\text{(Hole capacity, bbl/100 ft)} - \text{(Displacement \& capacity of drill pipe bbl/100 ft)}}$$

Reynolds Number Calculation (For Newtonian Fluids)

Reynolds number of 2,000 or less indicates laminar flow. A number of 4,000 or more indicates turbulent flow. Between the two is a zone of transition.

$$R_n = \frac{928\,DVW}{\mu}$$

where: D = diameter (inches) = hole
diameter minus drill pipe
O.D. (inches)
V = velocity (ft/sec) . . . average
flow velocity (see next equation)
W = mud weight (lb/gal)
μ = viscosity (cps)
R_n = Reynolds number

Critical Velocity (Non-Newtonian Fluids)

$$V_{ca} = \frac{3.8\sqrt{YP}}{W}$$

where: V_{ca} = critical velocity in annulus (ft/sec)
YP = yield point (lb/100 ft^2)
W = mudweight (lb/gal)

When the average flow velocity is greater than the critical velocity the annular flow is turbulent; if it is less, the flow is laminar.

Average Flow Velocity (Non-Newtonian Fluids)

$$V = \frac{Q}{2.448 \ (D_o{}^2 - D_i{}^2)}$$

where: V = average flow velocity (ft/sec)
        Q = flow rate (gpm)
        D_o = hole diameter (inches)
        D_i = drill-pipe diameter (inches)

Particle cuttings slip velocity

$$V = \frac{2G \ D^2(Ps - Pm)}{92.6 \ \mu}$$

where: V = drill cuttings slip velocity (ft/sec)
        G = gravity. . . . 32.2 ft/sec/sec
        Ps = specific gravity of cuttings $\times$ 62.4 lb/ft^3
        Pm = mud weight in lb/ft^3
           = mud weight in lb/gal $\times$ 7.48
        D = diameter of round particles or greatest
            diameter of flat cuttings *in feet*
$$\mu = \frac{(cps \ viscosity)(0.0672)}{100}$$

Annular pressure loss (Laminar Flow)

$$P = \frac{LYP}{225 \ D} + \frac{LPV(V)}{1500 \ D^2}$$

where: P = pressure drop (psi)
        L = drilling depth (ft)
     YP = yield point (lb/100 ft^2)
     PV = platic viscosity (cps)
        D = hole diameter minus drillpipe diameter (inches)
        V = mud velocity (ft/sec)

Bottom-hole circulating pressure (PSI)

BHCP = hydrostatic pressure + annular pressure loss

Equivalent circulating density (at the bit)(lb/gal)

$$ECD = \frac{BHCP}{Depth(ft) \times 0.052}$$

Bottomhole hydrostatic pressure reduction from gas-cut mud

$$\frac{\text{Weight of the gas cut mud}}{\text{Weight of uncut mud}} = \times\, p = n\; 2.3 \; \text{Log P}$$

where: p = pressure reduction from gas-cut mud (atmospheres)

$n = \dfrac{x}{1-x}$, Ratio of gas to mud

P = hydrostatic pressure in atmospheres

$\left(\text{Atmospheres} = \dfrac{\text{psi}}{14.7}\right)$

CALCULATIONS–GAS-PRODUCING WELL

Product	Equation	Units
Transmissibility	$Kh/\mu = 1{,}637\; Q_g ZT/m$	md-ft/cp
Theoretical flow capacity	$Kh = Kh\mu/\mu$	md-ft
Average effective permeability	$K = Kh/h$ $K_1 = Kh/h_1$	md
Indicated flow capacity	$(Kh)_2 = \dfrac{3{,}200\; Q_g\, \mu ZT\; \text{Log}(0.472\; b/r_{in})}{P_s^2 - P_f^2}$	md-ft
Damage ratio	$DR = \dfrac{\text{Theoretical flow cap}}{\text{Indicated flow cap}} = Kh/(Kh)_2$	–
Indicated flow rate	$OF_1 = \dfrac{Q_g\, P_s^2}{P_s^2 - P_f^2}$ Max.	Mcfd
	$OF_2 = \dfrac{Q_g\, P_s}{\sqrt{P_s^2 - P_f^2}}$ Min.	Mcfd
Theoretical potential rate	$OF_3 = OF_1 DR$ Max.	Mcfd
	$OF_4 = OF_2 DR$ Min.	Mcfd
Approx. radius of investigation	$b \approx \sqrt{Kt}$ or $\sqrt{Kt_o}$	ft
	$b_1 \approx \sqrt{K_1 t}$ or $\sqrt{K_1 t_o}$	ft
Potentiometric surface*	$\text{Pot.} = (El - GD) + (2.319\; P_s)$	ft

CALCULATIONS—LIQUID-PRODUCING WELL

Product	Equation	Units
Production	$Q = 1,440 \ R/t$	b/d
Transmissibility	$Kh/\mu = 162.6 \ Q/m$	md-ft/cp
Indicated flow capacity	$Kh = Kh\mu/\mu$	md-ft
Average effective permeability	$K = Kh/h$ $K_1 = Kh/h_1$	md md
Damage ratio	$DR = 0.183 \ (P_s - P_f)/m$	—
Theoretical potential with damage removed	$Q_1 = Q \ DR$	b/d
Approximate radius of investigation	$b \approx \sqrt{Kt}$ or $\sqrt{Kt_o}$ $b_1 \approx \sqrt{K_1 t}$ or $\sqrt{K_1 t_o}$	ft ft
Potentiometric surface*	$Pot. = El - GD + 2.319 \ P_s$	ft

(Courtesy of Halliburton Co.)

CALCULATIONS SYMBOLS

b = Approximate radius of investigation   ft
b_1 = Approximate radius of investigation (Net Pay zone h_1)   ft
D.R. = Damage ratio   —
El = Elevation   ft
GD = B.T. gauge depth (From surface ref.)   ft
h = Interval tested   ft
h_1 = Net pay thickness   ft
K = Permeability   md
K_1 = Permeability (from net pay zone, h_1)   md
m = Slope extrapolated pressure plot (psi^2/cycle gas)   psi/cycle
OF_1 = Maximum indicated flow rate   Mcfd
OF_2 = Minimum indicated flow rate   Mcfd
OF_3 = Theoretical open-flow potential with damage removed maximum   Mcfd
OF_4 = Theoretical open flow potential with damage removed minimum   Mcfd
P_s = Extrapolated static pressure   psig
P_f = Final flow pressure   psig
Pot = Potentiometric surface (fresh water*)   ft
Q = Average adjusted production rate during test   b/d
Q_1 = Theoretical production w/damaged removed   b/d

Q_g = Measured gas production rate   Mcfd
 R = Corrected recovery   bbl
r_w = Radius of well bore   ft
 t = Flow time   minutes
t_o = Total flow time   minutes
 T = Temperature, Rankine   °R
 Z = Compressibility factor   –
 μ = Viscosity gas or liquid   cp
Log = Common logarithms

* Potentiometric surface reference to rotary table When elevation not given, fresh water corrected to 100°F.

(Courtesy of Halliburton Co.)

Bibliography

API RP 40 (Aug 1960) "Recommended Practice for Core Analysis Procedure" Analysis Procedure" American Petroleum Institute. Copyright © 1960.

W. A. Bruce* & H. J. Welge* (23 May 1947) "The Restored State Method for Determination of Oil in Place and Connate Water." Production Practice and Technology.
  * The Carter Oil Co. Tulsa Okla. Paper presented at the Spring Meeting Mid-Continent District, Div. of Production, Amarillo Texas.

Core Laboratories Inc. "Fundamentals of Core Analysis" (No dates or other copyright data.)

W. J. Plumley, G. A. Risley, R. W. Graves Jr. and M. E. Kaley "Energy Index for Limestone Interpretation and Classification," Memoir 1, The American Association of Petroleum Geologists Edited by William E. Ham Copyright 1962.

M. W. Leighton and C. Pendexter "Carbonate Rock Types" AAPG as above.

W. E. Ham & L. C. Pray "Modern Concepts and Classifications of carbonate rocks. AAPG as above.

H. F. Nelson, C. W. Brown & J. H. Brineman. "Skeletal Limestone Classification. AAPG as above.

G. E. Thomas. "Textural and Porosity Units for Mapping Purposes." AAPG as above.

M. King Hubert, David G. Willis "mechanics of Hydraulic Fracturing" SPE-AIME Petroleum transactions Vol 210 Copyright 1957.

Richard B. Hohlt "The Nature & Origin of Limestone Porosity" Quarterly, Colorado School of Mines, Vol 43, #4. Copyright 1948.

A. I. Levorsen, Geology of Petroleum, Second Edition, W. H. Freeman and Company. Copyright © 1967.

Edward J. Lynch Formation Evaluation, First Edition, Harper and Row Copyright © 1962.

Graham B. Moody, Editor, Petroleum Exploration Handbook, First Edition, McGraw-Hill Book Company Inc. Copyright © 1961.

L. H. Robinson, "The Effect of Pore and Confining Pressure on the Failure Process in Sedimentary. Rock." Quarterly, Colorado School of Mines, Vol 54, copyright © 1959.

Harry M. Ryder, "Permeability, Absolute, Effective, Measured," World Oil, Copyright May 1948.

F. J. Pettijohn, Sedimentary Rocks, Second Edition, Harper and Row, Copyright © 1957.

Tool Pusher Manual, American Association of Oilwell Drilling Contractors, Seventh Edition, Manual prepared by Rotary Drilling Committee, December 1968. New manual called "The Drilling Manual" by International Association of Drilling Contractors, formerly AAODC.

Bureau of Standards Circular C. 410, 1936. "Reduction of Observed Degrees API to DEGREES API at 60°F.

Cecil R. Richardson* "Formation Testing Fundamentals," The Petroleum Engineer, 2 parts August & September 1960.
    * Halliburton Company, Duncan Oklahoma.

L. F. Maier,* "Recent Developments in the Interpretation and Application of DST Data," Journal of Petroleum Technology, Nov 1962.
    * Halliburton Oil Well Cementing Co., Ltd, Calgary, Alberta.

Dr. H. K. van Poollen & Sam J. Bateman,* "Application of DST to Hydrodynamic Studies," World Oil July 1958.
    * Halliburton Oil Well Cementing Company, Duncan, Okla.

W. C. Murphy,* "The Interpretation and Calculation of Formation Characteristics from Formation Test Data." Halliburton Company. Jan 1967.
    * Halliburton Company.

Halliburton Company, "Formation Testing Worldwide," June 1970.

Schlumberger, "Log Interpretation Principles," Schlumberger Ltd. Copyright USA 1969.

Schlumberger, "Log Interpretation Charts", Schlumberger Ltd. Copyright © 1969.

Christensen Diamond Products, "High Pressure Drop Diamond Bit Hydraulics," Christensen Diamond Products, SD 207F, January 1971.

Christensen Diamond Products, "Field Handbook," Christensen Diamond Products SD-206, January 1971.

Park J. Jones, "Mechanics of Production," Petroleum Production, Vol. 1, Reinhold Publishing Corporation.

Gruse & Stevens, The Chemical Technology of Petroleum, Second Edition (1942), McGraw-Hill Book Company.

Carlton Beal Jr., "The Viscosity of Air, Water, Natural Gas, Crude Oil, and its Associated Gases at Oil Field Temperatures and Pressures," Trans. American Institute of Mining and Metallurgical Engineers, Volume 165 (1946).

Sylvain J. Pirson, Oil Reservoir Engineering. Copyright 1958, by McGraw-Hill Book Company 2d Edition.

G. E. Archie, "Reservoir Rocks and Petrophysical Considerations." page 278, AAPG Bull., Vol. 36, #2, February 1952.

G. E. Archie, "The Electrical Resistivity log as an aid in determining some reservoir characteristics." Trans., AIME, Vol. 146 page 54, Copyright 1942.

W. L. Russell, "Well logging by radioactivity" AAPG Bull., Vol. 25, #9, page 1768, Sept 1941.

W. L. Russell, "The total gamma ray activity of sedimentary rocks as indicated by geiger counter determinations." Geophysics, Vol. 1X, #2, April 1944.

C. G. Brown, D. L. Katz, C. G. Oberfell, R. C. Alden, "Natural Gasoline and the Volatile Hydrocarbons," Gas Processors Association, formerly published by California Natural Gas Association, Bull TS-461, Sec. 1, page 44, 1948.

M. B. Standing, D. L. Katz, "Density of Natural Gases," SPE-AIME, formerly AIME, Trans., Vol. 146, page 140, 1942.

References

1. G. B. Moody, Petroleum Exploration Handbook, pge 18–1. Second Edition, McGraw-Hill Book Co.
2. A. I. Levorsen, Geology of Petroleum, pge 98. Second Edition, W. H. Freeman & Co.
3. Park J. Jones, "Mechanics of Production," Petroleum Production, Vol. 1 Reinhold Publishing Corp.
4. A. I. Levorsen, Geology of Petroleum, Second Edition page 98, W. H. Freeman & Co.
5. Core Laboratories Inc., The Fundamentals of Core Analysis, pages 8, 10, 11, 13.
6. F. J. Pettijohn, Sedimentary Rocks, Second Edition, page 73, Harper and Row.
7. M. King Hubert, David G. Willis, "Mechanics of Hydraulic Fracturing," AIME Petroleum Transactions, Vol. 210, pp 153–168.
8. L. H. Robinson, "The Effect of Pore and Confining Pressure on the Failure Process in Sedimentary Rock," Quarterly, Colorado School of Mines, Vol 54 (1959).
9. A. I. Levorsen, Geology of Petroleum, Second Edition, page 125, W. H. Freeman & Co.
10. Ibid page 121.
11. Harry M. Ryder, "Permeability, Absolute, Effective, Measured," World Oil, page 174, May 1948.
12. A. I. Levorsen, Geology of Petroleum, Second Edition, page 120, Figure 4–9, W. H. Freeman & Co.
13. Ibid page 106, Figure 4–1.
14. Core Laboratories Inc., The Fundamentals of Core Analysis, page 46.
15. API "Recommended Practice for Core Analysis Procedure," API RP 40, First Edition, August 1960, page 49, American Petroleum Institute.
16. Core Laboratories Inc., The Fundamentals of Core Analysis, page 55.
17. William E. Ham, Lloyd C. Pray, "Modern Concepts and Classifications of Carbonate Rocks," Memoir 1, Classification of Carbonate Rocks, page 6, 7. AAPG.
18. H. F. Nelson, C. W. Brown, J. H. Brineman, "Skeletal Limestone Classification," page 242, Memoir 1, Classification of Carbonate Rocks, AAPG.
19. G. E. Thomas, "Textural and Porosity Units for Mapping Purposes," Classification of Carbonate Rocks, Memoir 1, page 194, AAPG.
20. M. W. Leighton, C. Pendexter, "Carbonate Rock Types," Classification of Carbonate Rocks, Memoir 1, page 40, plate 3, AAPG.
21. Ibid page 37.
22. F. J. Pettijohn, Sedimentary Rocks, Second Edition, page 424, Harper and Row.
23. Ibid page 601.
24. Richard B. Hohlt, "The Nature and Origin of Limestone Porosity," Quarterly, Colorado School of Mines, Vol. 43 (1948), No. 4.

25. M. W. Leighton, C. Pendexter, "Carbonate Rock Types," Classification of Carbonate Rocks, Memoir 1, page 54, Plate VIII, AAPG.

26. R. W. Powers, "Arabian Upper Jurassic Carbonate Reservoir Rocks," Classification of Carbonate Rocks, Memoir 1, page 139–140, AAPG.

27. Ibid page 140.

28. W. J. Plumley, G. A. Risley, R. W. Graves Jr., M. E. Kaley, "Energy Index for Limestone Interpretation and Classification," Classification of Carbonate Rocks, Memoir 1, page 88–89, AAPG.

29. Ibid page 88, Table 1.

30. Ibid page 90, 92, 94, 96, 98. Plates I, II, III, IV, V.

31. Ibid page 97.

32. F. J. Pettijohn, Sedimentary Rocks, Second Edition, page 284, Harper and Row.

33. Ibid page 320.

34. Ibid page 298–299.

35. A. I. Levorsen, Geology of Petroleum, Second Edition, Page 438–440; W. H. Freeman & Co.

*36. Ibid page 422, Figure 10–4.

*37. Ibid page 436, Figure 10–11.

38. Ibid page 328.

39. Core Laboratories Inc., The Fundamentals of Core Analysis, pages 86, 87, Figures.

40. Ibid pages 83, 84, Figures.

41. A. I. Levorsen, Geology of Petroleum, Second Edition, page 537, W. H. Freeman & Co.

42. Ibid pages 296–298.

43. W. A. Bruce & H. J. Welge, "The Restored State Method for Determination of Oil in Place and Connate Water, page 170, Figure 9, Production Practice & Technology.

44. Ibid page 173, Figure 10.

45. A. I. Levorsen, Geology of Petroleum, Second Edition, page 477 (1st Ed. pge #) W. H. Freeman & Co.

46. Ibid page 481 (1st Ed. pge #).

*47. Ibid page 506, Table 11–2, After Claude E. Zobell's "Influence of Bacterial Activity on Source Sediments," Research on Occurrence & Recovery of Petroleum, American Petroleum Institute.

*48. Ibid. page 184, Table 5–6.

49. Gruse & Stevens, The Chemical Technology of Petroleum, Second Edition (1942), McGraw-Hill Book Company Inc.

50. A. I. Levorsen, Geology of Petroleum, Second Edition, page 185, W. H. Freeman and Company.

51. Ibid page 187.

52. Ibid page 342.

53. Carlton Beal Jr., "The Viscosity of Air, Water, Natural Gas, Crude Oil, and its Associated Gases at Oil Field Temperatures and Pressures," Transaction American Inst. Min. Met. Engrs, Vol. 165, (1946).

54. G. B. Moody, Petroleum Exploration Handbook, First Edition, page 18–1, McGraw-Hill Book Company Inc.

55. AAODC Tool Pusher's Manual, Seventh Edition AAODC, page 1, Section A-2. AAODC is now 1ODC . . . See bibliography. Ref. is Sec. A-5 page 2.

56. Ibid page 6, Section A-1.

57. AAODC Sec A-1 page 5 ammended by George Crosby, Div. of Engineering.

58. Ibid Sec A-1 page 5 (not ammended) (by G. Crosby).
59. RP40 "Recommended Practice for Core Analysis Procedure," pge 11, A.P.I.
60. Edward J. Lynch, Formation Evaluation, First edition, Harper & Row, page 200–201.
61. G. E. Archie, "The electrical resistivity log as an aid in determining some reservoir characteristics." Trans. SPE-AIME (formerly AIME), vol. 146, 1942, page 54.
62. Sylvain J. Pirson, Oil Reservoir Engineering. Copyright 1958, by McGraw-Hill Book Company, 2d Edition.
63. Schlumberger Ltd., Log Interpretation Charts. Por-1, C-10, page 14, 1969 Edition.
64. G. B. Moody, Petroleum Exploration Handbook, page 19–5, Second Edition, McGraw-Hill Book Company.
65. G. E. Archie, "Reservoir Rocks and petrophysical considerations." American Association of Petroleum Geologists, Bull. V.36, #2, Feb 1952, pge 278.
66. G. B. Moody, Petroleum Exploration Handbook, page 19–12, Second Edition, McGraw-Hill Book Company.
67. W. L. Russell, "Well logging by radioactivity," AAPG, Bull., V.25, #9, Sept 1941, page 1768.
68. W. L. Russell, "The total gamma-ray activity of sedimentary rocks as indicated by geiger counter determinations," Geophysics, Vol. IX, #2, April 1944.
69. G. B. Moody, Petroleum Exploration Handbook, page 20–13, figure 56, Second Edition, McGraw-Hill Book Company.
70. Schlumberger, Document #8 page 106, table 9 and Log Interpretation Charts, page 16., 1969 Edition.
71. Edward J. Lynch, Formation Evaluation, First edition, Harper and Row, page 279–280.
72. Schlumberger, "Log Interpretation Charts," Schlumberger Ltd. Copyright © 1969. page 29.
73. Ibid page 30.
74. Halliburton Company, "Formation Testing Worldwide" page 2.
75. Ibid page 3.
76. W. C. Murphy, "The interpretation and calculation of formation characteristics from formation test data," Halliburton Co., page 2, Fig. 1.
77. Ibid page 3, Fig. 2.
78. Ibid page 3, Fig. 3.
79. Ibid Page 4, Fig. 6.
80. Ibid Page 5, Fig. 7a.
81. Ibid Page 7, Fig. 10.
82. Ibid Page 6, Fig. 8.
83. Ibid Page 6, Figure 9.
84. Ibid Page 8, Fig. 13.
85. Ibid Page 8, Fig. 12.
86. Ibid Page 9, Fig. 15.
87. Ibid Page 9, Fig. 16.
88. Dr. H. K. van Poollen and Sam J. Bateman, "Application of DST to hydrodynamic studies," World Oil, July 1958.
89. L. F. Maier, "Recent Developments in the Interpretation and Application of DST Data, Journal of Petroleum Technology, Nov 1962. Page 1219.
90. Ibid Page 1219.
91. W. C. Murphy, "The Interpretation and Calculation of Formation Characteristics from Formation Test Data," Halliburton Company, page 19.

92. G. B. Moody, Petroleum Exploration Handbook, page 25–10, 25–11, Table 25–2, 25–3. Second Edition, McGraw-Hill Book Company.
93. Ibid page 25–20.
94. Ibid page 25–18.
95. C. G. Brown, D. L. Katz, C. G. Oberfell & R. C. Alden. Gas Processor's Association, originally published by California Natural Gas Association Bull. TS 461, Sec. 1 pge 44 1948, "Natural Gasoline and the Volatile Hydrocarbons."
96. M. B. Standing, D. L. Katz, "Density of Natural Gases," SPE-AIME, Formerly AIME, Trans., vol. 146, pge 140, 1942.
97. G. B. Moody, Petroleum Exploration Handbook, Appendix A, Table A-1, pge 2–3. Second Edition, McGraw-Hill Book Company.
98. Extract from Christensen Diamond Products Pamphlet SD 207F published January 1971.